Programmable logic devices and logic controllers

Enrique Mandado, Jorge Marcos
and Serafín A. Pérez
Institute for Applied Electronics
University of Vigo, Spain

PRENTICE HALL
London   New York   Toronto   Sydney   Tokyo   Singapore
Madrid   Mexico City   Munich

Originally published in Spanish as
Controladores lógicos y autómatas programables
by Marcombo SA, Gran Via de les Corts Catalanes, 594,
08007 Barcelona, Spain, © the authors, 1992.
First published in English (updated and revised) in 1996
from the Spanish second edition as *Programmable logic
devices and logic controllers* by
Prentice Hall Europe
Campus 400, Maylands Avenue, Hemel Hempstead,
Hertfordshire, HP2 7EZ
A division of
Simon & Schuster International Group

Spanish edition © the authors, 1992

English edition © Prentice Hall Europe, 1996

All rights reserved. No part of this publication may be
reproduced, stored in a retrieval system, or transmitted,
in any form, or by any means, electronic, mechanical,
photocopying, recording or otherwise, without the prior
permission, in writing, from the publisher. For permission
within the United States of America contact Prentice Hall
Inc., Englewood Cliffs, NJ 07632.

Typeset in Century and Times
by Mathematical Composition Setters Ltd, Salisbury

Printed and bound in Great Britain by
Redwood Books, Trowbridge, Wiltshire

Library of Congress Cataloging-in-Publication Data

Mandado, Enrique.
    [Controladores logicos y automatas programables. English]
    Programmable logic devices and logic controllers / Enrique
Mandado, Jorge Marcos, Serafín A. Pérez.
        p. cm.
    Translation of: Controladores logicos y automatas programables.
    Includes bibliographical references and index.
    ISBN 0-13-150749-4
    1. Programmable logic devices. 2. Programmable controllers.
I. Marcos, Enrique.   II Pérez, Serafín A.   III. Title.
TK7872.L64M36   1996
629.9'9—dc 20   95–53971
                                                              CIP

British Library Cataloguing in Publication Data

A catalogue record for this book is available from
the British Library

ISBN 0-13-150749-4

CONTENTS

Preface	xi
PART 1 LOGIC CONTROLLERS	**1**
1 Introduction to logic controllers	**3**
1.1 Preliminaries	3
1.2 Combinational logic controllers	4
1.3 Sequential logic controllers	7
1.3.1 Asynchronous logic controllers	8
1.3.2 Synchronous logic controllers	9
Specification of logic controllers	15
Non-modular synchronous logic controllers	25
Modular synchronous logic controllers	30
Semi-modular synchronous logic controllers	39
Bibliography	41
PART 2 LOGIC CONTROLLERS USING PROGRAMMABLE LOGIC DEVICES	**43**
2 Programmable logic devices	**45**
2.1 Introduction	45
2.2 Combinational PLDs	47
2.2.1 Universal combinational PLDs	47
Complete universal combinational PLDs	47
Incomplete universal combinational PLDs	52
2.3 Sequential PLDs	58
2.4 Advanced PAL-based PLDs	68
2.4.1 Fixed allocation advanced PAL-based PLDs	69
PLDs with one flip-flop and two feedback paths	69
PLDs with two flip-flops and two feedback paths	71

	2.4.2 Variable allocation advanced PAL-based PLDs	72
	PLDs with logic product steering	73
	PLDs with logic sum-of-products steering	74
	PLDs with multiple product term allocation	78
	PLDs with expander product term array	79
	2.4.3 PLDs with multiple arrays	80
2.5	PLDs using universal gates	84
2.6	PLD technologies	89
	Bibliography	89

3 Logic controller design using programmable logic devices — 91
3.1 Introduction — 91
3.2 Workstation for logic controller design using PLDs — 91
3.3 Logic controller design phases using PLDs — 92
3.3.1 Circuit capture — 92
3.3.2 Conversion to logic equations — 94
3.3.3 Logic equation minimization — 94
3.3.4 PLD selection — 95
3.3.5 Circuit behaviour simulation — 95
3.3.6 Programming file generation — 95
3.4 Logic controller design using PLDs — 95
3.4.1 Combinational logic controller design using PLDs — 95
3.4.2 Sequential logic controller design using PLDs — 104
Bibliography — 121

PART 3   PROGRAMMABLE LOGIC CONTROLLERS — 123

4 Introduction to programmable logic controllers — 125
4.1 Introduction — 125
4.2 Basic PLC with load and store instructions — 128
4.3 Basic PLC with conditional operating instructions — 133
Logic instructions — 137
Executive instructions — 139
4.4 Basic PLCs with improved characteristics — 141
4.4.1 PLC with input and output memory units — 141
4.5 Digital systems synthesis using PLCs — 143
4.5.1 Combinational system synthesis — 143
4.5.2 Edge-characterized sequential control system synthesis — 145
Programming transition capacity expressions — 146
Programming a transition graph — 149
4.5.3 Level-characterized sequential control system synthesis — 153
Bibliography — 155

5 PLC programming languages — 157
- 5.1 Introduction — 157
- 5.2 Instruction list — 158
  - 5.2.1 Variable identification — 158
  - 5.2.2 Instructions — 158
    - *Selection, input and output or operation instructions* — 158
    - *Timing and counting instructions* — 162
    - *Control instructions* — 165
- 5.3 Relay or ladder diagrams — 167
- 5.3.1 Variable identification — 167
  - 5.3.2 Logic sequences — 168
- 5.4 Function diagram — 172
  - 5.4.1 Variable identification — 172
  - 5.4.2 Logic operations — 172
- 5.5 Grafcet — 176
  - 5.5.1 Definitions and fundamental symbols of Grafcet — 176
  - 5.5.2 Application example — 178
- 5.6 Practical examples of digital system implementation using PLCs — 178
  - 5.6.1 Implementation of combinational systems using a PLC — 178
  - 5.6.2 Implementation of edge-characterized sequential control systems using a PLC — 181
  - 5.6.3 Implementation of level-characterized sequential control systems using a PLC — 183
- Bibliography — 203

6 Commercially available PLCs — 204
- 6.1 Introduction — 204
- 6.2 Simatic S5-100U — 204
  - 6.2.1 General characteristics — 205
  - 6.2.2 Programming — 206
    - *Variable identification* — 206
    - *Instruction set* — 207
  - 6.2.3 Practical examples of digital system implementation using the Simatic S5-100U — 224
    - *Combinational system implementation using the Simatic S5-100U* — 224
    - *Implementation of edge-characterized sequential control systems using the Simatic S5-100U* — 227
    - *Implementation of level-characterized sequential control systems using the Simatic S5-100U* — 232
- 6.3 Sysmac C-28K — 242
  - 6.3.1 General characteristics — 243

	6.3.2 Programming	244
	Variable identification	245
	Instruction set	247
	6.3.3 Practical examples of digital system implementation using the Sysmac C-28K	255
	Combinational system implementation using the Sysmac C-28K	255
	Implementation of edge-characterized sequential control systems using the Sysmac C-28K	258
	Implementation of level-characterized sequential control systems using the Sysmac C-28K	262

7 Microprocessor-based PLCs — 276
- 7.1 Introduction — 276
- 7.2 Microprocessors and control applications — 277
- 7.3 PLCs implemented using a microprocessor — 283
  - 7.3.1 General characteristics — 283
  - 7.3.2 Hardware and software resources — 285
- 7.4 Examples of PLCs with numerical processing capabilities — 288
  - 7.4.1 Simatic S5-100U — 288
    - *Hardware* — 288
    - *Software* — 290
  - 7.4.2 Sysmac C-28K — 297
    - *Hardware* — 297
    - *Software* — 298

8 Programming units and peripherals — 306
- 8.1 Introduction — 306
- 8.2 Programming units — 306
- 8.3 Peripheral units — 310

Appendix 1 Standard logic symbols — 313
- A1.1 Introduction — 313
- A1.2 Standard graphic symbol — 313
- A1.3 Symbols associated with inputs and outputs — 316
- A1.4 Standard representation of combinational systems — 320
  - A1.4.1 Generalities — 320
  - A1.4.2 Logic gates — 320
  - A1.4.3 Qualifying symbols for complex combinational functional blocks — 320
    - *AND dependency relationship [G(AND)]* — 323
    - *OR dependency relationship [V(OR)]* — 324
    - *Negation dependency relationship [N(NEGATE)]* — 325

        Enable/disable dependency relationship
           [EN(ENABLE)]   326
           Operating mode relationship [M(MODE)]   326
           Interconnection dependency relationship (Z)   327
           Address dependency relationship (A)   327
           Combination of dependency relationships   329
   A1.5 Sequential system standard representation   330
       A1.5.1 Introduction   330
       A1.5.2 Sequential system symbols   331
       A1.5.3 Sequential system dependency relationships   331
         Reset (R) and Set (S) dependency relationships   332
         Control dependency relationship (C)   332
         Operating mode dependency relationship (M)   334
   Combination of dependency relationships   335
       A1.5.4 Practical examples of sequential systems   339
         Parallel registers   339
         Counters   344
   Shift registers   349

Appendix 2 Real logic controller   **253**

Appendix 3 Commercial programmable logic devices   **356**
   A3.1 Introduction   356
   A3.2 Programmable logic device PLS100   356
   A3.3 Programmable logic device PLS155   358
   A3.4 Programmable logic device PLS157   361
   A3.5 Programmable logic device 5C031   362
   A3.6 Programmable logic device 5C060   365
   Bibliography   368

Index   **369**

PREFACE

Programmable logic devices (PLDs) and programmable logic controllers (PLCs) are becoming increasingly important in industrial applications. Because this development, motivated by the progress of microelectronics and programming techniques, has been very rapid, there are still a number of open questions with which even a specialist in applied electronics is often confronted and to which a straightforward answer has not been found.

What is the difference between a programmable logic controller and a computer? When should we use one or the other? Has the development of PLDs any influence on the applications of programmable logic controllers? Are they replacements for or complementary to programmable logic controllers?

This book answers these questions. By reading it, the reader will realize that computers, programmable logic controllers and programmable logic devices have in common the characteristic of being synchronous sequential systems, but he or she will also verify that they are different with respect to modularity, design confidentiality, speed, etc.

To benefit from this book the reader will need a good knowledge of basic digital electronics. Even so, the different items are explained in the text and up-to-date references are given at the end of each chapter.

The book is divided into three parts. The first consists of chapter 1 which introduces the reader to the topic of logic controllers and makes the connection between digital electronics and programmable logic controllers.

The second part, consisting of two chapters, is dedicated to programmable logic devices and their use in designing programmable logic controllers. In chapter 2, different types of programmable logic device and their characteristics are studied. Chapter 3 concentrates on the design of logic controllers using PLDs.

The third part is dedicated to programmable logic controllers (PLCs) and their applications. The basic concepts of PLCs and their main characteristics are introduced in chapter 4. In chapter 5, the different programming languages of PLCs are studied and their application is shown with various examples. Chapter 6 describes commercial PLCs and demonstrates their application using practical examples. Chapter 7 is dedicated to PLCs with computing capabilities. Their general characteristics are described and various commercial examples are analyzed. Finally, chapter 8 looks at programming units and their use in the design of PLC applications.

Appendix 1 describes the IEC standard logic symbols by means of numerous examples. It is included to help electronics engineers become familiar with the symbols that are now used to document new complex digital electronic circuit designs.

Enrique Mandado
Jorge Marcos
Serafín A. Peréz

PART 1
Logic controllers

In this part we study the different types of synchronous logic controller. These controllers are increasingly replacing their competitors owing to their inherent advantages and the continuing developments in the field of microelectronics.

CHAPTER 1

Introduction to logic controllers

1.1 Preliminaries

In many industrial applications it is necessary to implement process controllers that act upon one or more digital outputs as a function of the state, or the change of state, of some binary variables. A system that implements this function is called a **logic controller** because its inputs are binary variables.

Figure 1.1 shows how an industrial process is connected to a logic controller. By means of appropriate sensors, the process generates n binary variables. These variables are connected to the logic controller which, in turn, generates m output variables. Depending on how the output variables are used we have:

1. An open-loop control system if the output variables of the logic controller are only shown on a display to give information to the operator (Figure 1.2).
2. A closed-loop control system if the output variables of the controller are inputs to process actuators (Figure 1.3).

The logic controllers can be classified according to their operating mode as shown in Table 1.1. Each type of controller is described in the following sections.

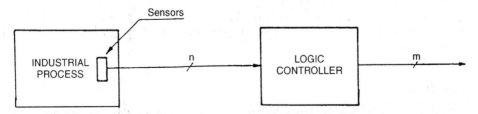

FIGURE 1.1 *Logic controller connected to an industrial process.*

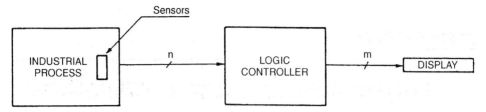

FIGURE 1.2 *Open-loop logic controller application.*

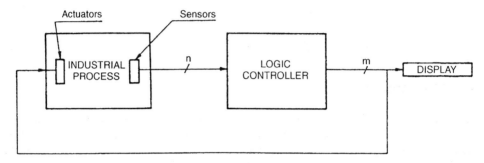

FIGURE 1.3 *Closed-loop logic controller application.*

TABLE 1.1 *Classification of logic controllers*

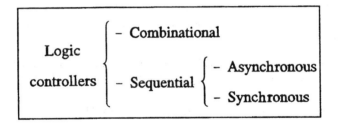

1.2 Combinational logic controllers

Combinational logic controllers have been the subject of much study in the digital electronics field [ALMA 94] [FLOY 94] [MAND 91] [WAKE 94]. In order to review the main problems of combinational design let us look at a simple example.

EXAMPLE 1.1

A chemical process has three temperature sensors at point P, whose outputs T_1, T_2 and T_3 assume two quite different voltage levels depending on whether the

INTRODUCTION TO LOGIC CONTROLLERS

temperature is less than or at least as great as t_1, t_2 or t_3, respectively ($t_1 < t_2 < t_3$). Zero is assigned to the voltage level corresponding to a temperature lower than t, and value 1 to the level corresponding to a temperature higher than or equal to t. Our purpose is to generate a binary signal that assumes the value 1 if the temperature lies between t_1 and t_2 or is greater than or equal to t_3, and the value 0 otherwise.

Solution

From the statement of the problem we conclude that a memoryless system is sufficient, since the value of the signal to be generated at a given time depends only on the logic level of the variables T_1, T_2 and T_3 at that time. Therefore, the circuit to be designed is a combinational system, as shown in Figure 1.4.

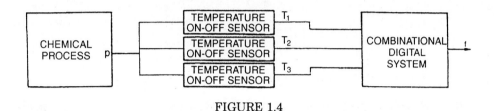

FIGURE 1.4

From the specifications we generate the corresponding logic table. Because the temperature cannot be simultaneously smaller than t_1 and higher than t_2 or t_3, the combinations of the input variables T_3, T_2, T_1, 010, 100, 101 and 110 cannot exist. The temperature of point P lies between t_1 and t_2 if $T_3 = T_2 = 0$ and $T_1 = 1$. Similarly, this temperature is higher than t_3 if $T_3 = T_2 = T_1 = 1$. Thus, output f must assume value 1 for these inputs only and for the rest, value 0. Hence the logic table of Table 1.2.

TABLE 1.2

Minterm	T_3	T_2	T_1	f
0	0	0	0	0
1	0	0	1	1
2	0	1	0	X
3	0	1	1	0
4	1	0	0	X
5	1	0	1	X
6	1	1	0	X
7	1	1	1	1

From this table we obtain the canonical expression of the sum of products form:

$$f = \sum_3 (1, 7) + \sum_0 (2, 4, 5, 6)$$

This expression leads to the Karnaugh map in Figure 1.5. If we assign value 1 to cell 5, we can then group cells 1 and 5 giving rise to the term 1-5, where variable T_3 is no longer present, thus resulting:

$$1\text{-}5 \equiv T_1 \overline{T_2}$$

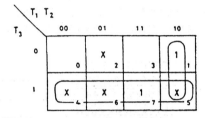

FIGURE 1.5 *Canonical product terms (minterms) Karnaugh map.*

If we also assign the logic value 1 to cells 6 and 4, we can implement the 1 in cell 7 by means of the group 4-5-6-7, where the variables T_1 and T_2 are no longer present, obtaining:

$$4\text{-}5\text{-}6\text{-}7 \equiv T_3$$

Therefore, the minimal expression for the sum of products is:

$$f = T_3 + T_1 \overline{T_2}$$

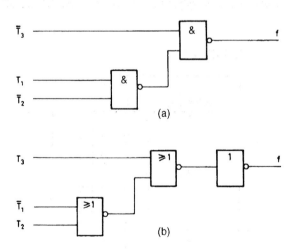

FIGURE 1.6 *Implementation of the function f. (a) Implementation using NAND gates. (b) Implementation using NOR gates.*

INTRODUCTION TO LOGIC CONTROLLERS

This expression can be implemented with NAND or NOR gates as follows:

$$f = \overline{\overline{T_3 + T_1 \overline{T_2}}} = \overline{\overline{T_3} \overline{T_1 \overline{T_2}}}$$

$$f = \overline{\overline{T_3 + T_1 \overline{T_2}}} = T_3 + \overline{\overline{T_1 \overline{T_2}}} = T_3 + \overline{T_1} + T_2$$

whose corresponding implementations are shown in Figure 1.6.

However, a combinational circuit is not a universal logic controller since it is not capable of making decisions that take into account the history of its inputs. Hence the interest in sequential logic controllers, which do have such capabilities.

1.3 Sequential logic controllers

As mentioned earlier, using a combinational logic controller it is not possible to implement a device whose outputs are functions of the sequences of transitions of its input variables. Therefore the need arises for systems capable of storing the sequences of input values, in the form of internal state, to make decisions as a function of the past values of their inputs.

Such systems are called **sequential systems** and the easiest way to implement them is with a feedback combinational system as shown in Figure 1.7. The memory of this sequential system is accomplished by the delay inherent to the combinational system.

We can also implement a sequential system with basic memory cells which are set to 0 or 1 through a combinational system as shown in Figure 1.8. Depending on the types of memory cell used, we have two different types of sequential system: asynchronous and synchronous.

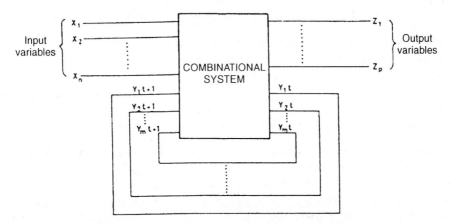

FIGURE 1.7 *Block diagram of a sequential system.*

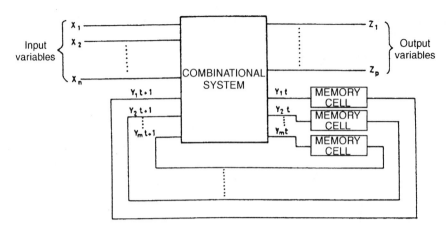

FIGURE 1.8 *Block diagram of a sequential system implemented using asynchronous memory elements.*

1.3.1 Asynchronous logic controllers

In asynchronous sequential systems the action of the input variables upon the system internal state is instantaneous. These systems can be implemented in two ways:

1. With a combinational system with direct feedback paths, as depicted in Figure 1.7.
2. With a combinational system with feedback through asynchronous memory cells, as in Figure 1.8.

When an asynchronous sequential system is used to control a given process, it becomes a logic controller (Figure 1.9).

Asynchronous logic controllers suffer from the following limitations:

1. Problems derived from hazards; these are very difficult to avoid when two or more inputs or internal state variables change simultaneously.
2. The impossibility of making the system modular, i.e. a change in the number of input or output variables is not possible without hardware changes. This statement is verified when we study methods to achieve modularity of synchronous logic controllers later in the book.

Nevertheless, asynchronous logic controllers have been thoroughly analyzed by various authors [ALMA 94] [McCL 86] [MAND 84] and used to implement simple control systems. On the other hand, developments in semi-custom digital integrated circuits and computer aided design techniques are continually reducing their field of application. The reader interested in asynchronous logic controllers is referred to the texts just mentioned.

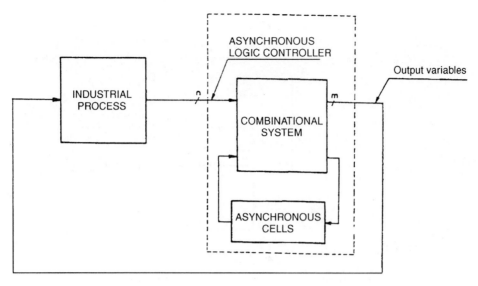

FIGURE 1.9 *Asynchronous logic controller*.

1.3.2 Synchronous logic controllers

From the above we can conclude that it is imperative to find new types of logic controller which exhibit the following characteristics:

1. Absence of hazards when several input or internal state variables change simultaneously.
2. The capability to change their functional specifications without hardware changes.
3. Input and output modularity.

These requirements can be accomplished by implementing a sequential system with the help of memory devices that have a control input that restricts the action of the input signals to well defined time instants. Such instants are defined by the edges of the system clock that is a square wave generated by an electronic device called a pulse generator. A simplified block diagram is shown in Figure 1.10. Synchronous binary devices are named **synchronous flip-flops** and a group of m flip-flops with a common clock input is called a **synchronous register**. In Figure 1.10, and in all of the remaining figures of this book, we adopt the standard symbols of the International Electrotechnical Commission (IEC). Since many readers may be unfamiliar with the new symbols, a description is provided in appendix 1.

How, though, can we avoid the disadvantages of asynchronous sequential systems mentioned in section 1.3.1?

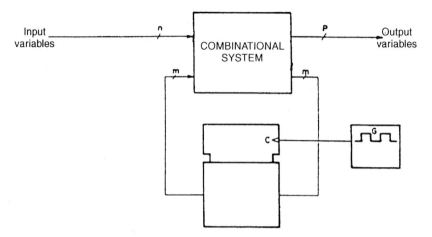

FIGURE 1.10 *Block diagram of a synchronous sequential system.*

The hazards, which arise when several input or internal state variables change simultaneously, are avoided if we ensure that the input variable is stable at the moment the active edges of the clock waveform are applied to the register of Figure 1.10. This can be done in two ways:

1. With an extra edge-triggered register placed between the input variables and the combinational circuit, and controlled by the same clock edges, as represented in Figure 1.11.

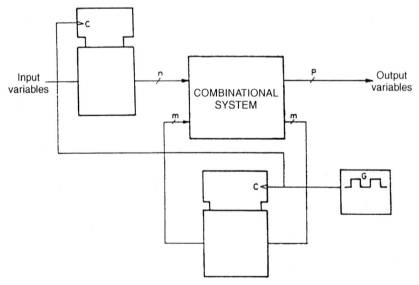

FIGURE 1.11 *Synchronous logic controller with an extra edge-triggered register to synchronize input variables.*

INTRODUCTION TO LOGIC CONTROLLERS   11

2. With two latch registers controlled by the same clock, as represented in Figure 1.12. Note that if the first register is sensitive to the high level of the clock signal, then the second will respond to the low level. The outputs of the first register are connected to the inputs of the second one. We thus ensure that the input variables will be stabilized when arriving at the combinational circuit.

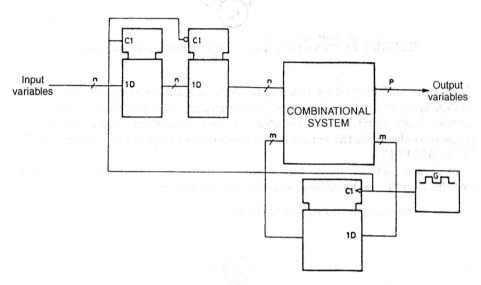

FIGURE 1.12 *Synchronous logic controller with two extra level-sensitive registers to synchronize input variables.*

The ability to change functional specifications without hardware changes is achieved using programmable combinational circuits (ROM, PROM, EPROM, PLA, PAL, etc.). This topic is so important that Part 2 of the book is dedicated to the subject.

The input and output modularity is obtained by looking at the inputs sequentially. This is possible because of the synchronous nature of the system. This subject is analyzed later in this section and is further studied in Part 3 of the book.

The use of a synchronous sequential system for implementing a process controller turns the system into a synchronous logic controller. Figures 1.11 and 1.12 show two possible implementations. In fact, the use of an input and output parallel register allows us to obtain a system with a state diagram as depicted in Figure 1.13, where we can jump between any two states. However, in Figure 1.13 some of the transitions occur between consecutive states. Furthermore, in the design of industrial logic controllers, which are the topic of chapter 3, we notice that to solve any digital control system a

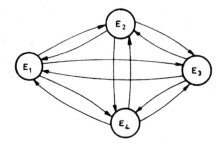

FIGURE 1.13 *State diagram of a synchronous sequential system.*

diagram as in Figure 1.14 is enough. Here, a decision is made to jump to a certain state or to the next state. In both cases it is possible to replace the internal state register by a counter, thereby simplifying the combinational system, as shown in the synthesis of synchronous sequential systems [ALMA 94] [MAND 91].

In Figure 1.15 we see a synchronous sequential system implemented with a counter and a combinational system. The counter has:

- m parallel input and output bits;

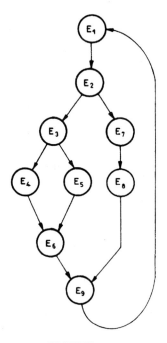

FIGURE 1.14

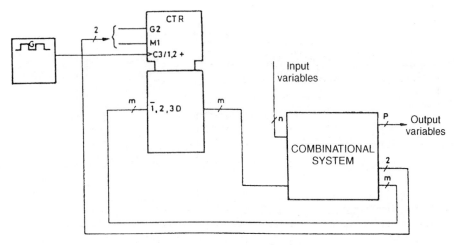

FIGURE 1.15 *Implementation of a synchronous sequential system using a counter and a combinational system.*

- an operation mode selection input M1 that activates the counting or the parallel input if it is 1 or 0, respectively;
- an inhibition input G2 which, when 0, inhibits the parallel information input or the counting (according to the state of M1).

The circuit in Figure 1.15 will change state with every active edge of the clock signal. Hence it can implement any state diagram, e.g. as in Figure 1.14.

A necessary condition for the correct operation of this circuit, and that of Figure 1.10, is that the clock period must be much smaller than the minimal interval between two consecutive input variable changes. Only if this condition is fulfilled can we guarantee that the system will respond to every input change.

In practice, the system in Figure 1.15 also exhibits the problems inherent to the lack of synchronism between the external variables and the clock signal. These become apparent when the system has to change state according to the simultaneous logic value of several external variables. To avoid the problem, we can synchronize the external input variables with the clock signal by means of the implementations shown in Figures 1.11 and 1.12.

In Figure 1.16 we show a synchronous logic controller implemented with a counter where the input variables are synchronized with a pair of latch registers.

From what has been said we can conclude that the systems in Figures 1.11, 1.12 and 1.16 base their decisions to activate the output variables on the state, or the sequence of states, of its input variables. This is why they

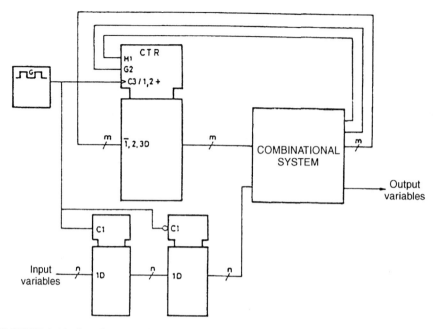

FIGURE 1.16 *Synchronous logic controller using a counter and synchronization of input variables by means of two extra level-sensitive registers.*

are simply called synchronous logic controllers. In practice we omit the word synchronous since it is implied.

To conclude this introduction, Table 1.3 lists the different ways of implementing logic controllers. The logic controllers of Figures 1.11, 1.12 and 1.16 are not modular because a change in the number of input or output variables implies a change in the system hardware. On the other hand, in

TABLE 1.3 *Classification of synchronous logic controllers*

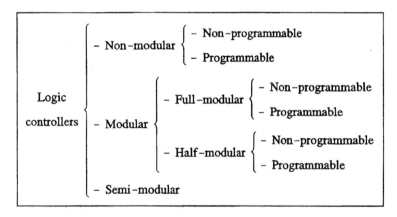

modular logic controllers we simply add elements to increase the number of input and output variables with no need to change the hardware.

Modular logic controllers allow the designer to achieve a technical solution of great flexibility, although at a much higher price than for the non-modular ones. Between these two approaches we have the semi-modular logic controllers, where the sum of the number of input and output variables remains constant but the number of each one can be modified. The different types of logic controller are analyzed next.

Specification of logic controllers

Logic controllers can be specified in two different ways, both of which have been thoroughly studied in the literature. They can be specified:

1. by input variable levels [HUFF 54] [UNGE 57] [McCL 65] [KOHA 70] [MAND 84], or
2. by means of edges or transitions of the input variables [MAND 76] [REY 74a] [REY 74b] [SMIT 71].

Defining logic controllers by input variable levels implies the analysis of all the states of the input variables for each internal state of the system. Thus, its practical application is difficult when the number of input variables, or their past strings to be memorized, is large. An example of controller specification by input variable levels is presented below.

EXAMPLE 1.2

We wish to implement an electronic lock with a non-modular logic controller specified by the levels of two binary variables A and B. When power is switched on the system is set at an initial state, from where the evolution of A and B is watched.

The lock must open if A and B are operated in the following sequence:

1. First A is activated and deactivated.
2. Then B is activated and deactivated.
3. Finally, A is activated and deactivated again.

If A and B are activated in the wrong sequence, the system returns to the initial state. It also returns to the initial state by the action of a microswitch M operated at the time the door is shut.

Solution

Figure 1.17 shows the state diagram of the specification by levels. When power is switched on the system is set at E_0 and remains there while A and B are 0. If A makes a transition from 0 to 1 the system changes to state E_1, from where it can:

- return to E_0 if B is set to 1;

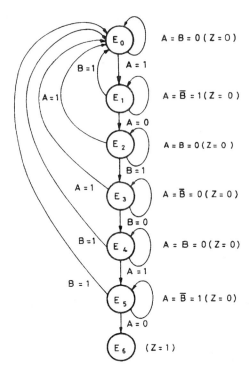

FIGURE 1.17 *State diagram of the level-sensitive logic controller of example 1.2.*

- change to E_2 if A returns to 0. In this state, the system remembers that A was activated.

Observing the system we can verify that it reaches E_4 if we activate and deactivate B only, and finally it reaches E_6 if A is again activated and deactivated.

This state diagram can be implemented with any of the logic controllers in Figures 1.11, 1.12 and 1.16 whose state is a function of the input variable levels.

Later in this chapter we obtain the diagram of a non-modular logic controller (Figure 1.28) which operates in the stated fashion and, in chapter 3, a semi-modular solution is presented.

Turning now to specification of logic controllers by means of edges or transitions of the input variables, in practice there are many sequential systems where the change of internal state must only take place when certain input variables change from 0 to 1, or vice versa, regardless of the state of the remaining variables. The specification by level changes is implemented by algebraic expressions of the transition capacity which is a binary variable that takes value 1 only when the appropriate input variable

INTRODUCTION TO LOGIC CONTROLLERS

changes are executed. The different expressions of the transition capacity are:

$$C_{T1} = \sum_{h=1}^{h=n} x_h \updownarrow (X_\beta^h)$$

$$C_{T2} = \sum x_h \updownarrow (X'_a)$$

$$C_{T3} = \sum x_h \updownarrow$$

$$C_{T4} = \sum X_a \updownarrow$$

These expressions specify the conditions that can induce the change of the internal state of an asynchronous sequential system:

- C_{T1} specifies that the change of the internal state must take place when a generic variable x_h changes state and all the others remain at a given state defined by X_β^h, where h varies from 1 to n.
- C_{T2} indicates that the internal state change must take place when x_h changes and a certain number of input variables remain at a given state.
- C_{T3} specifies the internal state changes caused by some input variable changes, regardless of the state of the others.
- C_{T4} corresponds to the internal state change due to the change of a vector of input variables.

A more detailed treatment of the specifications of edge-triggered asynchronous sequential systems can be found in [MAND 76] [MAND 91]. Nevertheless, the practical application of these methods can easily be understood, as shown in the examples below.

EXAMPLE 1.3

It is desired to move a cart C between points A and B which are indicated by two microswitches M_1 and M_2. The cart must be controlled by two push-buttons P_1 and P_2. It is initially parked at A and remains there until P_1 is pressed. At this time output Z_1 is activated, the cart motor is switched on and the cart starts moving towards B. This movement continues even if P_1 or P_2 is pressed. When the cart reaches point B it triggers microswitch M_2 which in turn activates variable Z_2 and deactivates variable Z_1, thus causing the cart to return to point A. If during this movement we press P_2, the cart should reverse direction, i.e. move again towards B. Consequently Z_1 must be activated and Z_2 deactivated again. If, on the other hand, P_2 is not pressed the cart continues towards A and it stops when microswitch M_1 is operated. In Figure 1.18 we depict the system and the block diagram of the logic controller.

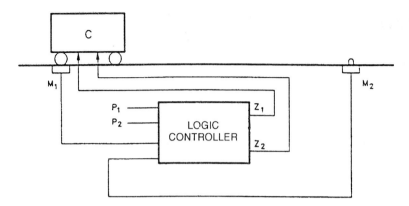

FIGURE 1.18 *Block diagram of the logic controller of a cart.*

Solution
In this example the decisions are taken whenever there is a change of state in any of the input variables, regardless of the state of the other variables. There are four input variables: push-buttons P_1 and P_2 and microswitches M_1 and M_2. The implementation of a primitive table of phases by the level method implies $2^4 = 16$ columns, which shows how hard it is to solve this problem using the said method.

Figure 1.19 represents the flowchart for the logic controller of the cart. This diagram can be directly obtained from the specifications as follows. When power is switched on the controller is set up automatically to its initial state, with outputs Z_1 and Z_2 deactivated. While P_1 does not change from 0 to 1 the controller remains in its initial state, regardless of any changes that might occur in the remaining variables.

When P_1 eventually changes from 0 to 1, the controller changes its internal state, and output variable Z_1 is activated. The action of P_1 on the programmable controller must be independent of the time it remains pressed; therefore it is the change from 0 to 1 that activates the controller. In this case we have $C_T = P_1\uparrow$.

From this moment, the controller must only watch to see if the cart arrives at its final position, which takes place when microswitch M_2 changes from 0 to 1. The transition capacity is, therefore, $C_T = M_2\uparrow$, which is indicated with the corresponding symbol in Figure 1.19. Immediately after a transition of M_2 is detected, Z_1 is deactivated, Z_2 is activated and the cart reverses.

From now on, the logic controller must watch for the changes in push-button P_2 and microswitch M_1. Consequently the algebraic expression for C_T is $C_T = P_2\uparrow + M_1\uparrow$, which is indicated by two appropriate symbols in Figure 1.19.

If P_2 commutes, Z_2 is deactivated and Z_1 is activated, and the cart moves again towards M_2. Consequently the transition capacity C_T is the same as after P_1 is activated and the logic controller returns to the same transition state. If, on the contrary, M_1 is activated indicating that the cart has reached the end of the track, Z_2 must be deactivated in order to stop it, and the controller returns to the start-up state.

INTRODUCTION TO LOGIC CONTROLLERS 19

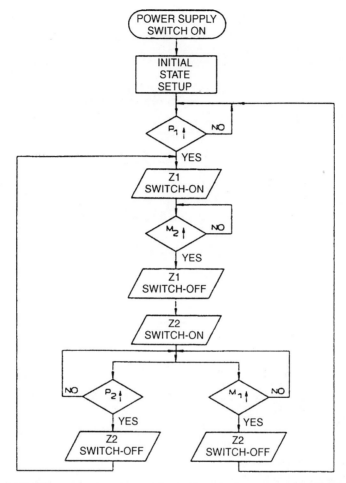

FIGURE 1.19 *Flowchart of the logic controller of Figure 1.18.*

Figure 1.20 shows the flowchart of Figure 1.19 indicating explicitly the internal transition states. From this we obtain the state diagram represented in Figure 1.21.

EXAMPLE 1.4

The final product of a production line is metal bars whose length must not exceed L. To select the final product the system of Figure 1.22 is used, which consists of a conveyor belt carrying the bars through two photoelectric detectors located a distance L apart. These detectors consist of a light emitter and receiver. The output of the light receivers adopts two distinct voltage levels according to the presence or absence

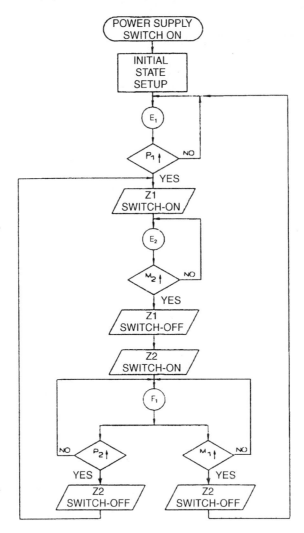

FIGURE 1.20 *Flowchart of the logic controller of Figure 1.18 indicating explicitly the internal transition states.*

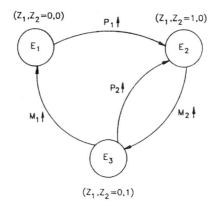

FIGURE 1.21 *State diagram of the logic controller of Figure 1.18.*

INTRODUCTION TO LOGIC CONTROLLERS   21

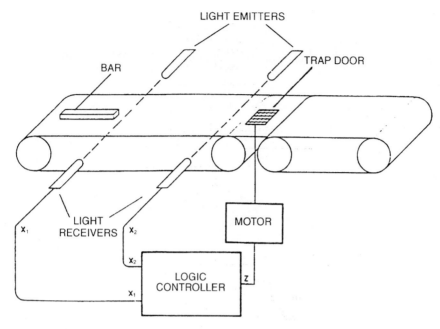

FIGURE 1.22 *Bar selection system in accordance with its length.*

of a bar between them and the emitter. By convention we assign value 1 to the output when the bar is in front of the detector and 0 otherwise.

After the second detector there is a trapdoor activated by a motor M. If the bar has a length greater than L, M must be excited to open the trap door and let the bar drop. Otherwise M is not excited. Once the bar has dropped, motor M is deactivated and returns to its normal state and the system is ready for a new detection.

The distance that separates the bars to be tested must be such that only one bar at a time enters the inspection zone.

The problem is to design a logic controller whose inputs are the outputs of the detectors x_1 and x_2 and whose output Z activates motor M.

Solution
From the problem statement we conclude that, from an initial state, a bar must be rejected when x_1 is in state 1 at the time that x_2 changes from 0 to 1. Hence the flowchart of Figure 1.23. From the initial conditions the controller detects if there is a change of x_2 from 0 to 1 and simultaneously if x_1 is 1, which indicates that the bar has length greater than L. The transition capacity has, therefore, the expression $C_T = x_2 \uparrow (x_1 = 1)$ which, when equal to 1, activates the output Z.

The deactivation of Z can take place after x_2 changes from 1 to 0.

In Figure 1.24 we represent the flowchart of Figure 1.23 but indicating explicitly the internal transition states. From this we obtain the state diagram of Figure 1.25.

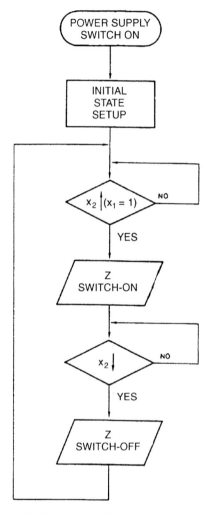

FIGURE 1.23 *Flowchart of the logic controller of the bar selection system of Figure 1.22.*

EXAMPLE 1.5

We wish to design a logic controller specified by level changes that implements the control of the electronic lock of example 1.2.

Solution

This example is less suited to implementation by edges than are the two previous examples because in each state it is necessary to watch and make a decision as a function of all the changes of the input variables. The corresponding state diagram is presented in Figure 1.26.

INTRODUCTION TO LOGIC CONTROLLERS

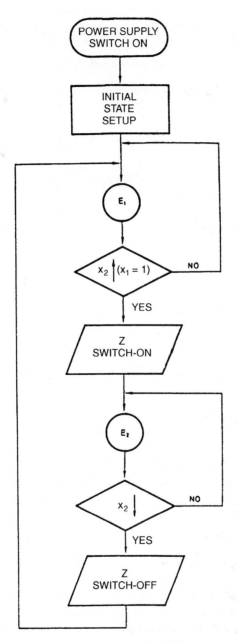

FIGURE 1.24 *Flowchart indicating explicitly internal transition states.*

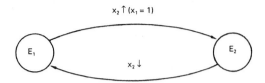

FIGURE 1.25 *State diagram of the logic controller of the bar selection system of Figure 1.22.*

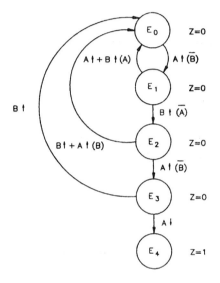

FIGURE 1.26 *State diagram of the edge-triggered logic controller of the lock of example 1.5.*

The controller is initially in state E_0 and can only change to E_1 if A is activated and B not. This is detected when the transition capacity $C_T = A\uparrow(\overline{B})$ assumes the value 1. The system should change to state E_2 if, subsequently, B is activated after A has returned to 0. This is detected when $C_T = B\uparrow(\overline{A})$ assumes the value one. Any other change must induce the transition E_1 to E_0. Consequently, the transition capacity from E_1 to E_0 is $C_T = A\downarrow + B\uparrow(A)$.

The reader is invited to complete the remaining part of Figure 1.26.

In order to implement any of the flowcharts obtained in the above examples with a synchronous logic controller it is necessary to convert the edges into logic operations. Fortunately the periodic sampling of the input variables allows the specification of level changes using logic products. In fact, an input variable changes state when its value at sampling instant t is different from the value at the previous sampling instant $t-1$. As a consequence, level changes can be expressed algebraically as:

$$x_i\uparrow \equiv x_{it}\overline{x_{it-1}} = 1$$
$$x_i\downarrow \equiv \overline{x_{it}}x_{it-1} = 1$$
$$x_i\updownarrow \equiv x_{it}\overline{x_{it-1}} + \overline{x_{it}}x_{it-1} = 1$$

Any of the state diagrams of Figures 1.21, 1.25 and 1.26 can be transformed with the help of the expressions above. Therefore the diagrams in Figures 1.25 and 1.27 are equivalent.

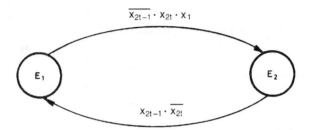

FIGURE 1.27 *State diagram of the logic controller of Figure 1.22 indicating level changes by means of logic products.*

The logic controllers of Figures 1.11, 1.12 and 1.16 cannot implement the state diagrams specified by edges obtained in the previous examples. Therefore Figure 1.28 shows a non-modular synchronous logic controller specially designed to implement state diagrams specified by edges. This system has an internal state register R_E, with feedback through a programmable universal combinational system (PUCS). The outputs of registers R_1 and R_2 are connected to the PUCS's input. The R_1 register receives the p input variables x_i and generates the p synchronized input variables x_{it}. In turn, these are subdivided into two groups of p' and p'' variables, which act by edges and by levels, respectively. The R_2 register receives in its input the p' variables which act by edges and generate in its output the delayed variables x_{it-1}.

The output variables of both registers are connected to the combinational circuit, which computes the different transition capabilities. In chapter 3 we present several examples of logic controller implementation based on this method.

Non-modular synchronous logic controllers

Non-modular synchronous logic controllers, which can be specified either by levels (Figures 1.11, 1.12 and 1.16) or by edges (Figure 1.28), have the

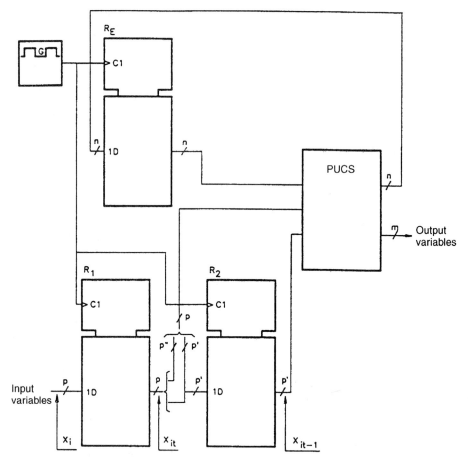

FIGURE 1.28 *Diagram of a non-modular synchronous logic controller to implement state diagrams specified by edges.*

following characteristics:

1. Their hardware can be minimized by adapting it to the needs of the application.
2. The combinational circuit, which is responsible for the decision process, can be implemented using either:
   - interconnected logic gates giving rise to a wired logic controller, or
   - ROMs, PROMs, RPROMs, PLAs or PALs. In this way, we obtain a programmable logic controller whose behaviour can be modified without hardware changes.

INTRODUCTION TO LOGIC CONTROLLERS

3. It is possible to implement the whole system with a single custom or semi-custom integrated circuit, thereby protecting the copyrights of the circuit and the machine it controls.

4. They do not easily allow expansions because an increase in the number of inputs or outputs which was not planned in advance generally requires changes in the combinational circuit.

From these characteristics we conclude that this type of controller is suitable whenever it is required to minimize costs and we do not expect an increase in the number of input or output variables.

Wired logic controllers are difficult to design without the help of a computer, particularly when the number of inputs plus internal states is greater than six. As confirmation, we present below the design of a wired logic controller.

EXAMPLE 1.6

Design a non-modular synchronous wired logic controller defined by levels that implements the control system of the electronic lock of example 1.2.

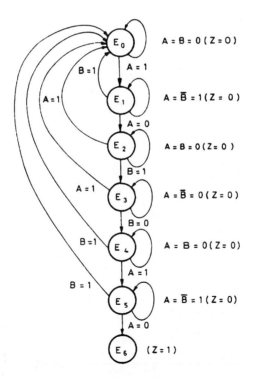

FIGURE 1.29

Solution

In Figure 1.29 we present again the state diagram obtained in example 1.2. From it we derive the following steps:

1. The functional block that stores the internal state of the controller is selected, which can be a 3-bit natural binary synchronous counter (in practice, if the system is implemented on medium-scale integration circuits, we can use a 4-bit BCD natural or binary natural counter, because of its availability as functional block integrated in TTL or CMOS).

2. From the state diagram we obtain the logic table shown in Table 1.4. In this table we see that if the Xs of the D_0, D_1 and D_2 variables are converted into 0, the corresponding logic equations are:

$$D_0 = D_1 = D_2 = 0$$

As shown in Figure 1.29, this happens because we must jump to 0 from every state.

3. From Table 1.4 we obtain the Karnaugh maps for I and $C/\overline{P}$ shown in Figure 1.30

TABLE 1.4

			t			$t+1$					
B	A	Q_2	Q_1	Q_0	Q_2	Q_1	Q_0	I	$C/\overline{P}$	D_2	D_1	D_0
X	0	0	0	0	0	0	0	1	X	X	X	X
0	1	0	0	0	0	0	1	0	1	X	X	X
1	1	0	0	0	0	0	0	1	X	X	X	X
0	0	0	0	1	0	1	0	0	1	X	X	X
0	1	0	0	1	0	0	1	1	X	X	X	X
1	X	0	0	1	0	0	0	0	0	0	0	0
0	0	0	1	0	0	1	0	1	X	X	X	X
X	1	0	1	0	0	0	0	0	0	0	0	0
1	0	0	1	0	0	1	1	0	1	X	X	X
0	0	0	1	1	1	0	0	0	1	X	X	X
X	1	0	1	1	0	0	0	0	0	0	0	0
1	0	0	1	1	0	1	1	1	X	X	X	X
0	0	1	0	0	1	0	0	1	X	X	X	X
1	X	1	0	0	0	0	0	0	0	0	0	0
0	1	1	0	0	1	0	1	0	1	X	X	X
1	X	1	0	1	0	0	0	0	0	0	0	0
0	1	1	0	1	1	0	1	1	X	X	X	X
0	0	1	0	1	1	1	0	0	1	X	X	X
X	X	1	1	0	1	1	0	1	X	X	X	X
X	X	1	1	1	X	X	X	X	X	X	X	X

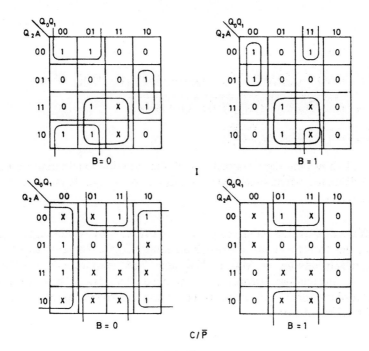

FIGURE 1.30

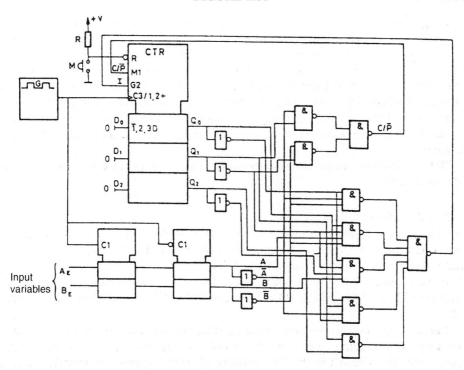

FIGURE 1.31

and finally the equations:

$$I = \overline{Q_0\overline{A}\overline{B} + Q_0\overline{Q_1}A\overline{B} + Q_0\overline{Q_1}\overline{Q_2}B + Q_0\overline{Q_1}\overline{A}B + Q_1Q_2}$$

$$= \overline{\overline{Q_0\overline{A}\overline{B}}\,\overline{Q_0\overline{Q_1}A\overline{B}}\,\overline{Q_0\overline{Q_1}\overline{Q_2}B}\,\overline{Q_0\overline{Q_1}\overline{A}B}\,\overline{Q_1Q_2}}$$

$$C/\overline{P} = Q_1\overline{A} + \overline{Q_1}B = \overline{\overline{Q_1\overline{A}}\,\overline{\overline{Q_1}B}}$$

Figure 1.31 shows the logic controller with the combinational circuit implemented with NAND gates. Microswitch M is connected to the reset R input of the counter (enabled with 0).

The difficulties in the design and implementation of wired logic controllers are overcome by the use of programmable logic devices together with computer aided design (CAD) techniques. The importance of these procedures is increasing rapidly, motivated by the fast progress of microelectronics, and it is for this reason that the second part of this book is dedicated to their study.

Modular synchronous logic controllers

The logic controllers studied in the previous section are non-modular because all the input variables are connected to different terminals of the combinational circuit, as are all the output variables. However, there are many applications where the need to change the number of input and output variables arises. A typical example is an industrial plant that undergoes changes during its lifetime. In these cases, it is better to use a logic controller that has a modular input and output structure which allows us to adapt the number of modules to the needs of the application without changing the hardware.

Before studying modular logic controllers it is useful to consider the issue of modularity.

INPUT AND OUTPUT MODULARITY

A synchronous logic controller can achieve modularity of input variables by connecting the variables to a single line or bus and selecting them sequentially, so that the state of only one variable appears on the bus at any time. The bus concept has been made possible by the development of tri-state circuits [ALMA 94] [FLOY 94] [McCL 86] [MAND 91] [PROS 87] [WAKE 94].

Figure 1.32 shows the corresponding diagram with 2^n input variables and an n variable decoder. Each input variable is connected to a tri-state output gate with an enable input. The outputs of these gates are connected to a single wire (1 bit bus).

INTRODUCTION TO LOGIC CONTROLLERS 31

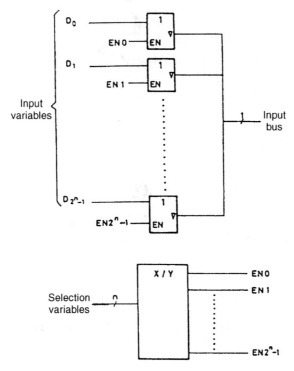

FIGURE 1.32 *Input unit connected to a bus.*

The enable input of each gate is connected to a different output of the decoder. In this way, only the input variable corresponding to the combination applied to the inputs of the decoder appears on the bus at any one time.

To achieve modularity, the 2^n input variables are grouped into 2^m subgroups, or modules, of 2^{n-m} variables each. Moreover, it must be possible to connect the modules and to ensure that only one module is selected at a time. To this end, the m variables of a group are compared with the combination produced by a set of switches and the output of this comparator is connected to the enable (EN) input of the decoder, which generates the enable variables of the above-mentioned gates (Figure 1.33).

Figure 1.34 shows the block diagram of the module of Figure 1.33. Figure 1.35 represents the modular input unit formed by the 2^{n-m} internally programmed modules in order to define the groups 0 to $2^{n-m} - 1$.

The modularity of the output variables can be achieved by connecting them to type D flip-flops, with the data inputs connected to a single output bus (Figure 1.36). These flip-flops are selected sequentially; each has an enable input (G1) which is connected to an output of a decoder whose inputs are the selection variables.

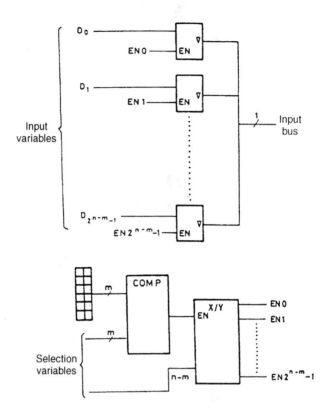

FIGURE 1.33 *Input variables module*.

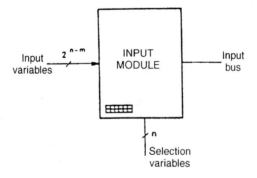

FIGURE 1.34 *Block diagram of an input variables module*.

INTRODUCTION TO LOGIC CONTROLLERS 33

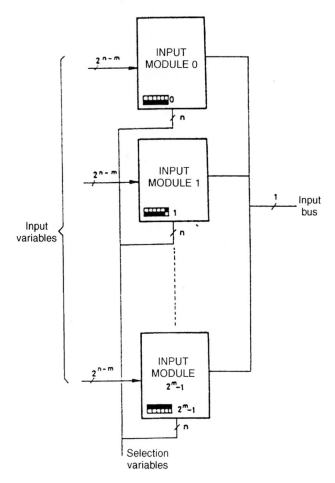

FIGURE 1.35 *Modular input unit.*

The 2^n output variables can be grouped into 2^m subgroups of 2^{n-m} variables each, in the same way as the input variables, hence obtaining the output module represented in Figure 1.37 and the modular output unit of Figure 1.38.

IMPLEMENTATION OF MODULAR LOGIC CONTROLLERS
Combining the set of input and output modules described in the previous section with the synchronous sequential system of Figure 1.15, we obtain the modular logic controller shown in Figure 1.39.

Comparing Figures 1.39 and 1.16 we conclude that no synchronization registers for the input variables are necessary because the system watches

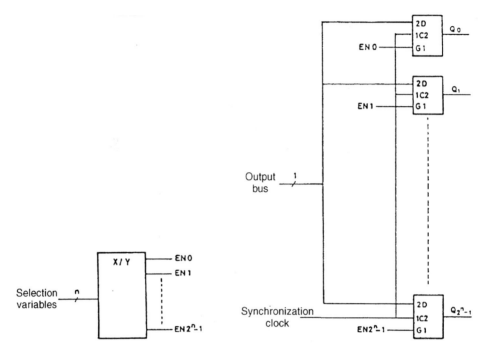

FIGURE 1.36 *Output unit connected to a bus.*

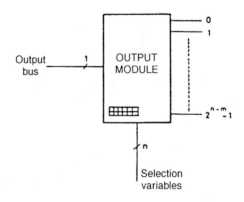

FIGURE 1.37 *Block diagram of a 2^{n-m} output variables module.*

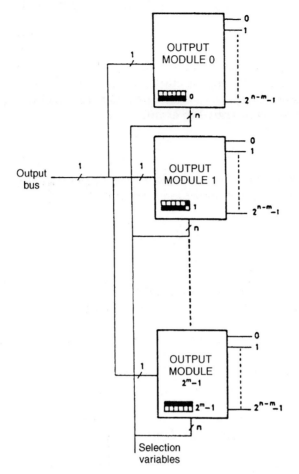

FIGURE 1.38 *Modular output unit.*

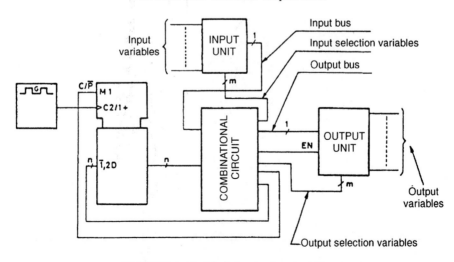

FIGURE 1.39 *Modular logic controller.*

only a single input variable at each clock cycle. This fact prevents the logic controller of Figure 1.39 implementing a logic OR or AND of input variables, or a combination of both, in a single internal state. However, it can be done sequentially, as demonstrated in example 1.7.

EXAMPLE 1.7

Explain how the following logic functions of the input variables a, b and c can be implemented with a logic controller as in Figure 1.39.

$f_1 = abc$

$f_2 = ab + c$

Solution

Given that only one input variable is applied to the combinational circuit at any time, it is necessary for the system to take decisions about successive jumps according to the logic state of each variable.

Figure 1.40 represents the state diagram of the system that implements the function $f_1 = abc$. By simple inspection the reader will confirm that the programmable controller reaches E_3 if any of the a, b or c variables is 0, and that it reaches E_4 only if all variables are equal to 1.

Figure 1.41 represents the state diagram of the system that implements the function $f_2 = ab + c$. The reader can confirm that the system reaches state E_4 only when f_2 is 0 and that it reaches state E_3 when f_2 is 1.

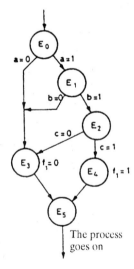

FIGURE 1.40 *State diagram of the system that implements the function $f_1 = abc$.*

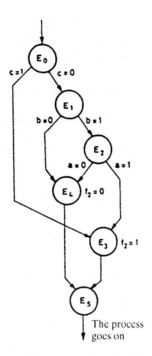

FIGURE 1.41 *State diagram of the system that implements the function* $f_2 = ab + c$.

From what has just been said, we can conclude that the system of Figure 1.39 can perform any logic function through a sequence of decisions. In order to be able to set the output flip-flops to 1 or 0 at the right time, the combinational circuit generates an enable variable (EN) such that only when EN equals 1 will the information present on the output bus be transferred to the selected flip-flop. The output of the combinational circuit of Figure 1.39 is represented in the vector of Figure 1.42.

One of the most important aspects of the system of Figure 1.39 is, undoubtedly, the way of implementing the combinational circuit, which determines the state diagram of the controller, and, consequently, its behaviour. This implementation can be achieved in two ways:

1. By means of a set of **interconnected gates**. In this way we obtain a modular wired logic controller, which requires a change in the combinational circuit whenever a change in its behaviour is required. Therefore, if

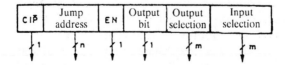

FIGURE 1.42 *Output vector of the combinational circuit of Figure 1.39.*

the combinational circuit is implemented with gates, the advantages of modularity are lost since an increase in the number of input or output variables will require hardware changes. The immediate consequence, then, is that although wired logic controllers are theoretically possible, their implementation has no practical application.

2. By means of a **programmable combinational circuit**. The implementation of the logic controller of Figure 1.39 by means of a programmable combinational circuit allows functional alterations without hardware changes.

However, a detailed analysis of the circuit of Figure 1.39 reveals the following characteristics:

1. The number of bits of the vector of Figure 1.42 is excessively large. In fact, in the controller of Figure 1.39 the selection of input and output variables is achieved by means of independent variables. This allows simultaneously an output action and a decision as a function of an input variable. The length of the instruction can be reduced if a single field is used for the input and output and a bit is allocated to the selection of the action to be performed.

2. The output of the input unit is directly applied to the combinational circuit. Hence, whenever specifications are changed, the input of the combinational circuit also changes, as does its logic table.

Therefore, an interesting alternative to the logic controller of Figure 1.39 is the one of Figure 1.43 in which the output of the input unit is connected to a three-channel multiplexer, whose output is in turn connected to the M1 input of the counter. The other two channels of the multiplexer are held at logic levels 0 and 1, and the field $C/\overline{P}$ of the vector of Figure 1.42 is replaced by the bits S_0 and S_1. At the output of the multiplexer we have the state of the input bus, the level 0 (jump of the counter) or the level 1 (counting of the counter), if S_0 and S_1 are 00, 01 or 10, respectively.

Moreover, the logic controller of Figure 1.43 presents a single selection field, shared by the input and the output units and an input/output operation mode variable I/O, which enables the input or output unit when it assumes the value 0 or 1, respectively. The output vector of the combinational circuit is therefore as shown in Figure 1.44 (*cf.* Figure 1.42). The improvements introduced show that there are many ways of designing logic controllers of the type just analyzed [BOUT 76] [ZSOM 83]. In appendix 2 we present a brief description of the logic controller developed at McGill University [ZSOM 83] which combines the selection of output variables with the jump address in order to reduce the number of bits of the output vector of the combinational circuit.

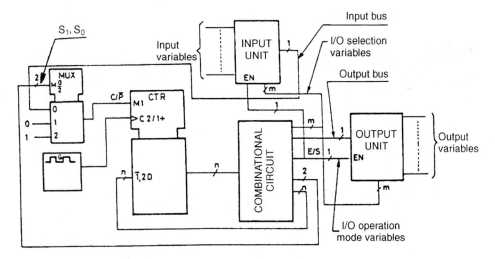

FIGURE 1.43 *Programmable logic controller with input and output single selection.*

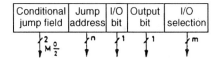

FIGURE 1.44 *Output vector of the combinational circuit of Figure 1.43.*

It is also of interest to remark that the output vectors of the combinational circuit of the controllers of Figures 1.39 and 1.43 shown in Figures 1.42 and 1.44, respectively, give no clear indication about the operation performed by the controller because they do not have any field of operation code. This led to the conversion of these controllers into programmable logic controllers (PLCs), which are studied in Part 3 of this book.

Semi-modular synchronous logic controllers

Semi-modular synchronous logic controllers combine the reduced hardware complexity of non-modular controllers with the flexibility of modular ones. Their basic diagram is shown in Figure 1.45, where we can see that the internal state is stored in register R_1. R_1's input variables can be divided into two groups:

1. n variables which are connected to a similar number of output variables of the combinational circuit, and which constitute the minimum number of internal state variables.

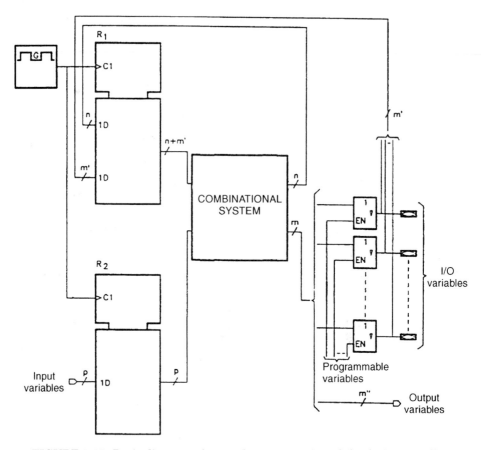

FIGURE 1.45 *Basic diagram of a synchronous semi-modular logic controller.*

2. m' variables which are connected to a similar number of output variables of the combinational circuit via tri-state gates. These in turn are connected to output pins.

The tri-state gates have independent enable inputs which can be set to 0 or 1. When an EN variable is 1, the output of the corresponding gate is not in the third state and the binary variable assigned to it can be either:

- an output variable, which appears at the output pin to which the gate is connected, or
- an internal state variable. In this case the corresponding output pin is not used.

Conversely, if the EN variable is 0, the output of the gate is set in the third state and an input variable can be connected to the corresponding pin.

From this we conclude that these logic controllers fulfil the semi-modularity condition defined at the beginning of this section.

Semi-modular logic controllers are characterized by not using, in general, the entire supporting hardware since in most of the applications there are idle R_1 register flip-flops and tri-state input gates.

The digital system of Figure 1.45 is a functional block implementable on an integrated circuit. Therefore it is of interest to semiconductor manufacturers to offer these systems as standard off-the-shelf integrated circuits called programmable logic devices (PLDs). This type of circuit has been made possible only by the progress of integration techniques; however, the development of these techniques has been so fast that a general theory which can systematize their analysis is required. Such a theory is expounded in chapter 2 and constitutes a systematic method of describing PLDs, which simplifies their use in the design of logic controllers (the topic of chapter 3).

Bibliography

[ALMA 94] A.E.A. Almaini, *Electronic Logic Systems*, 3rd edition. Prentice Hall, 1994.
[BOUT 76] R.T. Boute, 'The binary machine as a programmable controller', *Euromicro Newsletter*, vol. 1, no. 2, 1976.
[FLOY 94] T.L. Floyd, *Digital Fundamentals*, 5th edition. Prentice Hall, 1994.
[HUFF 54] D.A. Huffman, *The Synthesis of Sequential Switching Circuits*, no. 257, 1954.
[KOHA 70] Z. Kohavi, *Switching and Finite Automata Theory*. McGraw-Hill, 1970.
[MAND 76] E. Mandado, 'Nuevos métodos sistemáticos de síntesis de autómatas asíncronos de control', PhD thesis, ETSI Industriales de Barcelona (Spain), 1976.
[MAND 84] E. Mandado, *Sistemas electrónicos digitales*, 5th edition. Marcombo, 1984.
[MAND 91] E. Mandado, *Sistemas electrónicos digitales*, 7th edition. Marcombo, 1991.
[McCL 65] E.J. McCluskey, *Introduction to the Theory of Switching Circuits*. McGraw-Hill, 1965.
[McCL 86] E.J. McCluskey, *Logic Design Principles*. Prentice Hall, 1986.
[PROS 87] F.P. Prosser and D.E. Ninkel, *The Art of Digital Design*. Prentice Hall, 1987.
[REY 74a] A.C. Rey and J. Vaucher, *Self-synchronous Control*, 7th Annual Workshop on Microprogramming. ACM, 1974.
[REY 74b] A.C. Rey and J. Vaucher, 'Self-synchronized asynchronous sequential machines', *IEEE Transactions on Computers*, December 1974.
[SMIT 71] J.R. Smith and J.R. Roth, 'Analysis and synthesis of asynchronous sequential networks using edge-sensitive flip-flops', *IEEE Transactions on Computers*, August 1971.
[UNGE 57] S.H. Unger, 'A study of asynchronous logical feedback networks', PhD thesis, Department of Electrical Engineering, Massachusetts Institute of Technology, 1957.

[WAKE 94]   J.F. Wakerly, *Digital Design*, 2nd edition. Prentice Hall, 1994.
[ZSOM 83]   P.J. Zsombor-Murray, L.J. Vroomen, R.D. Hudson, T. Le-Ngoc and P. Holck, 'Binary decision based programmable controllers', *IEEE Micro*, vol. 3, 1983.

PART 2
Logic controllers using programmable logic devices

In chapter 1, logic controllers and their various implementations were analyzed. This second part of the book is dedicated to the design of logic controllers using programmable logic devices (PLDs), which constitute the best solution to the design of semi-modular logic controllers.

CHAPTER 2

Programmable logic devices

2.1 Introduction

Programmable logic devices (PLDs) may be defined as logic circuits implemented on a single integrated circuit, with the capability to be programmed to implement any type of combinational and/or sequential system.

These circuits can be implemented in large (LSI) or in very large (VLSI) scale integration and they have the following advantages over logic gates and standard sequential circuits (counters or registers) implemented in small (SSI) or medium (MSI) scale integration:

1. They allow the implementation of a whole digital system in a single integrated circuit instead of having to use several interconnected circuits.
2. They reduce the complexity of the printed circuit that supports the digital system.
3. As a consequence, they show better reliability and noise immunity, shorter propagation times and less power dissipation.
4. Because of their programming facility, we can reconfigure the design without changing the hardware.
5. It is impossible to copy the controller, therefore the copyright of the design is protected.

From the above arguments we conclude that PLDs are standard, off-the-shelf integrated circuits suitable for solving a specific application by means of appropriate hardware programming. Since the initial circuit is independent of the application, and during the design it is configured to the needs of the particular application, PLDs are also known as configurable integrated circuits.

There are several types of programmable logic device which can be characterized by their logic function, as shown in Table 2.1. They are studied in subsequent sections.

TABLE 2.1 *PLDs characterized by their logic function*

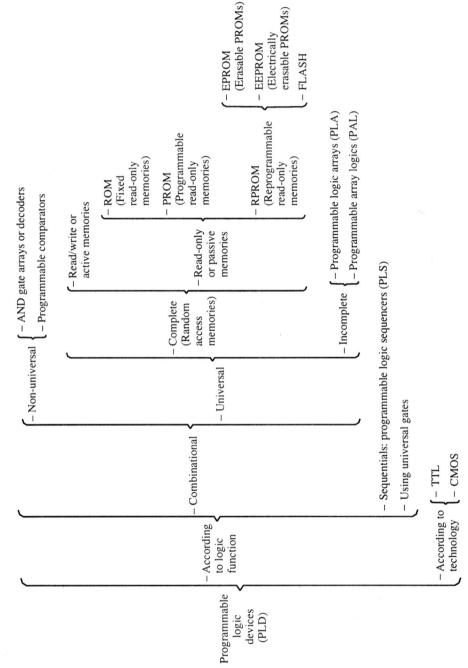

2.2 Combinational PLDs

Combinational programmable logic devices are combinational circuits whose logic table can be modified without changing the hardware, simply by eliminating (programming) certain connections.

Programmable logic devices may be universal or non-universal. The former can be used to implement any logic function, while the latter can only implement certain functions of general application. The most important non-universal programmable logic devices are AND gate arrays (or decoders) and programmable comparators. Their applications fall outside the context of this book and we do not deal with them here. The interested reader is referred to [TEXA 89] [PHIL 87].

Universal combinational PLDs are the most important or essential building blocks of synchronous logic controllers; they are the topic of the next section.

2.2.1 Universal combinational PLDs

When the value of the output variables for each combination of input variables can be programmed independently then PLDs are called **complete** devices. Otherwise they are called **incomplete**.

Complete universal combinational PLDs

Complete combinational systems are defined as those where it is possible to program independently the value of the output variables for each combination of input variables. Random access memories, in their different versions, are complete universal combinational programmable logic devices. A random access memory (RAM) consists of N cells, capable of storing binary data (0 or 1) grouped in positions of m cells, such that the total number p of positions verifies the equation $N = pm$. The memory has, in the most general case, m input pins whose data can be transferred to the m cells of any position in a single write operation; it has also m output pins capable of receiving the data of the m cells of any position in a single read operation. Both sets of pins can be regarded as a single group used to transfer data to or from the memory. Figure 2.1 represents the block diagram of a random access memory (RAM). The reader not familiar with the standard symbols should consult appendix 1.

To select the position to be read or written, the memory has n address pins such that $2^n = p$. Each of the 2^n possible combinations of the n address variables selects one of the p memory positions.

The reader can easily understand that a random access memory acts as a combinational system during the reading operation, because for each binary combination present at the n address inputs, we get an output equal to the contents of the selected position which is independent of the input vector sequences. The memory address inputs are the input variables of the

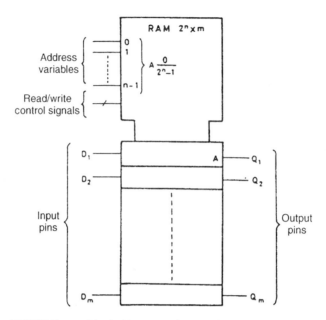

FIGURE 2.1 *Block diagram of a random access memory.*

combinational system and the information outputs are the output variables, as indicated in Figure 2.2. Furthermore, this combinational system is programmable because the data present in each position can, via writing operations, be replaced with the data present at the input pins. Hence, the input pins are the programming inputs of the circuit (Figure 2.2). The read/write control signal allows the selection of reading or writing operations as a function of its logic level.

Therefore, a random access memory acts as a combinational system during a reading operation, as indicated graphically in Figure 2.3. In this figure, the read/write control signal is not represented because it is assumed to be held at the logic level corresponding to the reading operation. The input pins are used exclusively to do the programming by writing operations in the different memory positions.

Random access memories can be classified, as indicated in Table 2.1, as read/write or active and read only or passive. Read/write memories are volatile, i.e. they lose information when the supply voltage is switched off. For this reason we can use them as programmable combinational systems only if they have a very low power consumption in order to allow battery-powered operation. Consequently, the most frequently used are the CMOS technology memories.

Read-only or passive memories are not volatile and are therefore widely used as programmable combinational systems. They are classified as fixed

PROGRAMMABLE LOGIC DEVICES

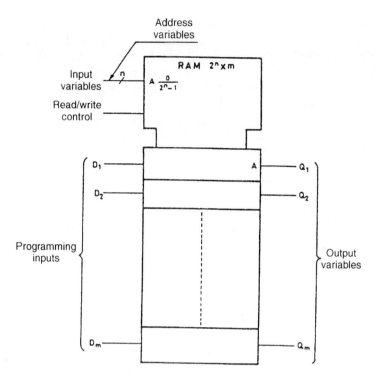

FIGURE 2.2 *Random access memory as a programmable combinational system.*

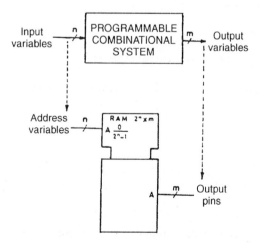

FIGURE 2.3 *Logic symbol of a programmable combinational system implementation using a random access memory.*

read-only memories (ROMs), programmable read-only memories (PROMs) and reprogrammable read only memories (RPROMs). Their differences reside mainly in the way they are programmed. In particular, the RPROM's memories can be EPROMs (electrically programmable read-only memories), EEPROMs (electrically erasable programmable read-only memories) or FLASH. For a more detailed study the reader is referred to [ALMA 94] [FLOY 94] [PROS 87] [MAND 91] [WAKE 94].

The above-mentioned memories can have a tri-state output. Figure 2.4 represents two logic symbols of a ROM of 2^n positions of m bits with a tri-state output.

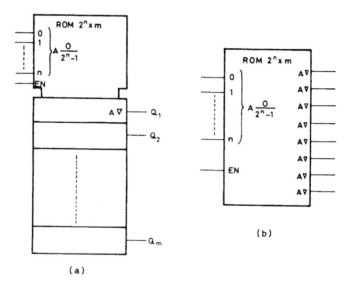

FIGURE 2.4 *Standard logic symbols of a fixed read only-memory (ROM) of 2^n addresses × m bits with tri-state output.*

Any combinational circuit can be implemented with a read-only or passive memory as shown in the following example.

EXAMPLE 2.1

Implement with a PROM the combinational system described in example 1.1 whose logic table is shown in Table 2.2.

Solution
First it is necessary to determine memory organization, i.e. the number of positions and the number of bits for each position.

The number of positions has to be equal to the number of combinations of the input variables, i.e. $2^3 = 8$. Each position must have one bit corresponding to the

TABLE 2.2

T_3	T_2	T_1	f
0	0	0	0
0	0	1	1
0	1	0	X
0	1	1	0
1	0	0	X
1	0	1	X
1	1	0	X
1	1	1	1

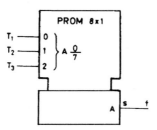

FIGURE 2.5

output variable f. The memory is represented in Figure 2.5. The input variables T_1, T_2 and T_3 are connected to the address inputs A_0, A_1 and A_2 (A 0/7), respectively, and the output pin S coincides with the output variable f.

In each of the memory locations we store the information about f corresponding to each input variable combination indicated in Table 2.2.

This example shows clearly that passive memories are misused when they are employed to generate logic functions because there are memory positions whose contents become irrelevant. On the other hand, passive memories are the most adequate programmable logic devices to store the instructions of a modular programmable logic controller, such as the one represented in Figure 1.39, because each counter combination is associated with a microinstruction.

To be able to compare passive memories with the incomplete programmable logic devices to be studied in the next section, it is useful to analyze the implementation of a passive memory using a two-dimensional array by means of an AND gate array connected to an OR gate array, as represented in Figure 2.6.

The number of gates of the AND gate array equals the number of combinations of the input variables, i.e. 2^n, and each AND gate decodes a canonical product. For this reason, it is a non-programmable array. The fixed connections are represented by a dot (●).

The number of gates of the OR gate array equals the number of variables in each position. Each OR gate has 2^n inputs, each one connected to the output of an AND gate by connections that can be eliminated, as shown by the Xs in Figure 2.6. Programming consists in eliminating those connections of the inputs of an OR gate corresponding to a memory position for which the output is 0.

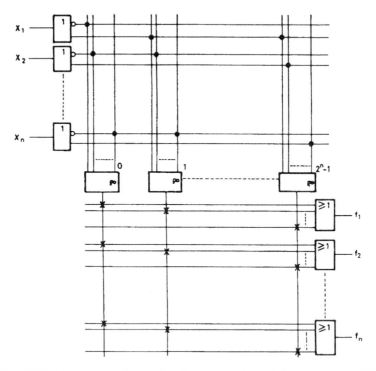

FIGURE 2.6 *Diagram of a read-only memory (ROM, PROM or RPROM) implemented by means of an AND gate array and an OR gate array.*

Therefore, a passive memory can be considered as a fixed AND gate array and a programmable OR gate array. To simplify the diagram, the OR and AND gate inputs can be represented by just one connection, as indicated in Figure 2.7, where we explicitly mention that the AND gate array is really a decoder.

Incomplete universal combinational PLDs

The complete PLDs discussed above are able to program independently the value of the output variables for each of the 2^n possible combinations of the n input variables. However, in practice, the following situations

PROGRAMMABLE LOGIC DEVICES 53

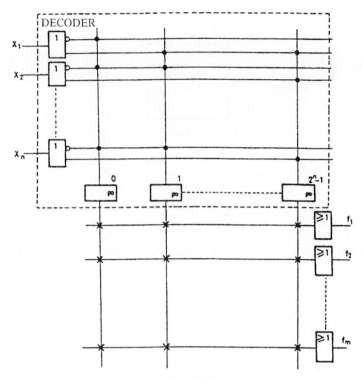

FIGURE 2.7 *Simplified diagram of the read-only memory of Figure 2.6.*

may arise:

1. The logic functions are at logic 1 level only for a number of input variable combinations less than 2^n.
2. The canonical expression of the sum-of-products can be simplified by numerical methods or by Karnaugh tables.
3. The function is not defined for some combinations of the input variables.

None of these three cases can be simplified when the function is implemented by means of a complete combinational programmable system; hence the advantages in using incomplete programmable combinational systems, so-called because it is not possible to program independently the value of each output variable for each combination of the input variables. Their wide applicability is based on the fact that, in practice, almost all of the functions can be simplified and can be represented by a minimal non-canonical algebraic expression as studied in digital electronics [ALMA 94] [FLOY 94] [McCL 86] [MAND 91] [WAKE 94].

Incomplete universal PLDs consist of an AND gate programmable array connected to a set of OR gates (Figure 2.8) allowing the direct

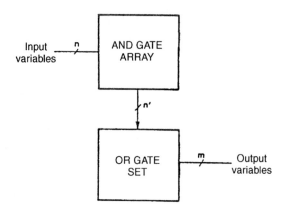

FIGURE 2.8 *Block diagram of an incomplete universal combinational logic device.*

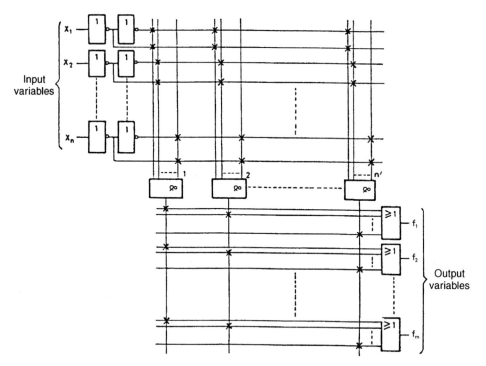

FIGURE 2.9 *Diagram of a programmable logic array.*

PROGRAMMABLE LOGIC DEVICES

implementation of any logic function expressed by a sum of logic product.

According to the way the OR and the AND gates are connected, two different kinds of incomplete programmable logic device are possible: PLAs and PALs. These are dealt with below.

PROGRAMMABLE LOGIC ARRAYS

A programmable logic array (PLA) consists of an AND gate programmable array with n' gates ($n' < 2^n$) and a programmable array of m OR gates. The AND gates have n inputs connected to each input variable and to its inverse via a connection that can be eliminated (Figure 2.9).

The programming of such a logic array consists in suppressing both the appropriate connections of the AND gate array in such a way that the output of each AND gate represents a certain logic product, and also the required connections of the OR gate array so that each output is the sum of the appropriate AND gate outputs.

The diagram of Figure 2.9 can be represented in a compact form, as indicated in Figure 2.10.

To condense the diagram of a digital system with a PLA of the type shown in Figures 2.9 and 2.10, we can use the block diagram of Figure 2.11, where

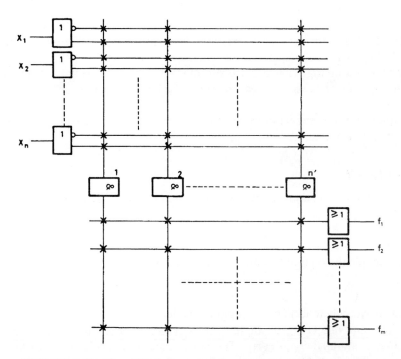

FIGURE 2.10 *Simplified diagram of a programmable logic array.*

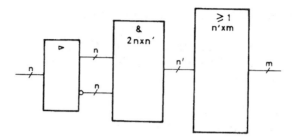

FIGURE 2.11 *Block diagram of a programmable logic array.*

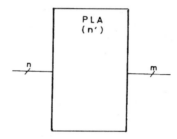

FIGURE 2.12 *Simplified block diagram of a programmable logic array.*

the AND gate array and the OR gate array are indicated separately. An even simpler block diagram is shown in Figure 2.12, with only one logic block labelled PLA and n', where n' denotes the number of AND gates which are part of the array.

Programmable logic arrays are the most flexible incomplete PLDs because it is possible to program the connection of each product to any of the output OR gates.

A simple implementation example of a logic function with a programmable logic array is analyzed below.

EXAMPLE 2.2

Implement the combinational system described in example 1.1 using a programmable logic array.

Solution
In this example the minimal expression for function f is:

$$f = T_3 + T_1 \overline{T_2}$$

From it we conclude that the programmable logic array must have at least two AND gates and one OR gate. Figure 2.13 represents the simplified diagram of the programmed minimal logic array. The reader can confirm that it is much simpler than the passive memory system of example 2.1.

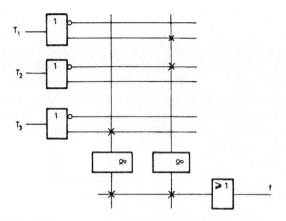

FIGURE 2.13 *Minimum PLA to implement the combinational system of example 2.2.*

PROGRAMMABLE ARRAY LOGICS

Programmable array logics (PALs) correspond to the block diagram of Figure 2.8 and they differ from programmable logic arrays (PLAs) by the fact that the OR gate inputs are connected directly to a given number of AND gates. In general, if the PAL has n' products and m outputs, each OR gate is connected to n'/m different products. Figure 2.14 shows a PAL with twelve logic products and three OR gates with each of them connected to four products.

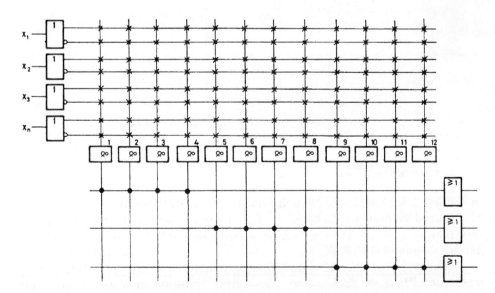

FIGURE 2.14 *PAL with twelve logic products and three output variables.*

PALs are less flexible than PLAs, because if a logic product has to be part of two outputs, it has to be programmed twice. On the other hand, PALs have a shorter propagation time, less power dissipation and, above all, require a smaller area of silicon as their programming is simpler. In most cases, given a PAL with a number of inputs, products and outputs, we can implement the same functions as with a PLA of similar complexity. The system described in example 2.2 above confirms this. For all these reasons, PALs are becoming increasingly important in the design of digital systems with programmable logic circuits.

Figure 2.15 represents the block diagram of a programmable array logic (PAL) with n' products and m OR gates, each with n'/m inputs, and in Figure 2.16 we have the condensed block diagram.

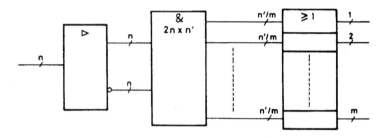

FIGURE 2.15 *Block diagram of a PAL.*

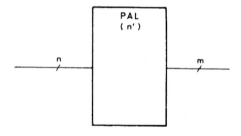

FIGURE 2.16 *Simplified block diagram of a PAL.*

2.3 Sequential PLDs

In section 1.3.2 we studied synchronous logic controllers and saw that they can be used to implement any combinational or sequential digital system. For this reason, interest grew in implementing sequential programmable logic devices (sequential PLDs).

Section 2.2 describes the first programmable logic devices (PLDs): they were just PLAs or PALs, and those from different manufacturers were almost identical. However, the increase of integration density led to the

PROGRAMMABLE LOGIC DEVICES

combination of PLAs and PALs with synchronous flip-flops, to produce programmable logic sequencers (PLSs).

Figure 2.17 shows the basic structure of a PAL-based PLS that consists of the following elements:

- a programmable interconnect matrix or array (PIA) of m input variables;

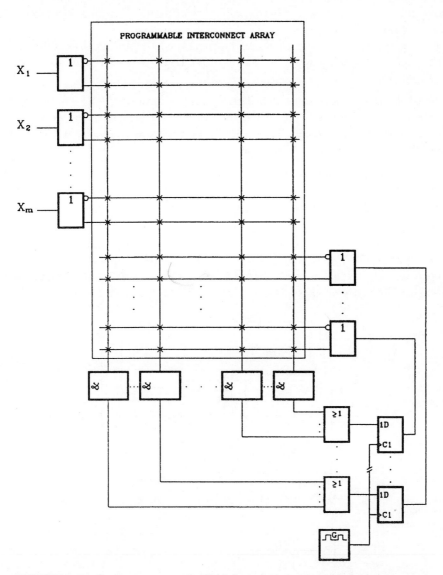

FIGURE 2.17 *Basic structure of a PAL-based programmable logic sequencer.*

- n' AND gates;
- a number p of OR gates, each connected to the output of n'/p AND gates;
- a number p of edge-triggered D flip-flops. The output Q of every flip-flop is fed back to the matrix.

The PLS can also be represented as in Figure 2.18. The programmable interconnect matrix is represented as a functional block called 'programmable interconnect array' (PIA) and all the outputs Q of the flip-flops are output variables.

The performance of the PLS can be improved by adding different circuits to the basic structure shown in Figure 2.18. Examples of these additions are as follows:

1. Tri-state gates between each flip-flop Q output and the corresponding external pin (Figure 2.19). Each one of these tri-state gates is controlled by an enable input (EN). This allows the external pin to be used as input or output, depending on the state of EN.

2. Set (S) and Reset (R) inputs to the flip-flop. As shown in Figure 2.20, it is possible to establish the internal state at power-on if R and S inputs are added to every flip-flop.

3. Inverting or non inverting outputs, by placing a two-input exclusive OR gate between the Q output of every flip-flop and the corresponding external pin (Figure 2.21). This provides programmable output polarity because the inversion control input (N1) selects whether the external pin level is the same as Q or $\overline{Q}$.

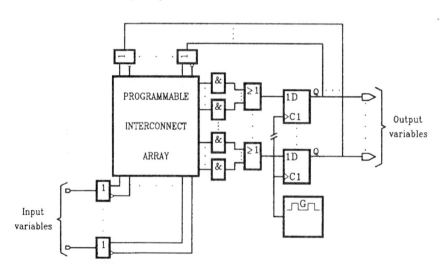

FIGURE 2.18 *Basic structure of a PAL-based PLS.*

PROGRAMMABLE LOGIC DEVICES

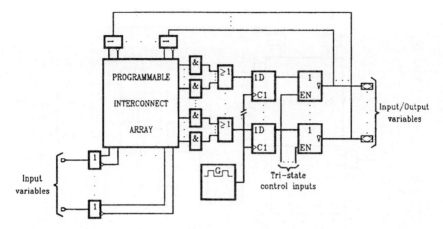

FIGURE 2.19 *Basic PAL-based PLS with tri-state gates.*

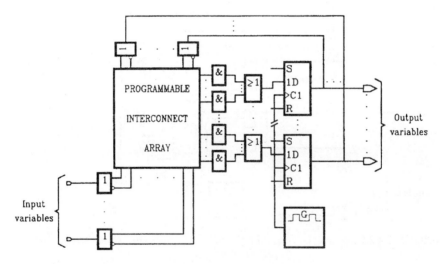

FIGURE 2.20 *Basic PAL-based PLS with flip-flop set and reset inputs.*

4. Multiplexing of the OR gate and the flip-flop outputs. A two-channel multiplexer is included in Figure 2.22. One of its inputs is connected to the OR gate output, and the other to the flip-flop Q output. The OR gate output appears at the external pin if the multiplexer selection input is high, whereas the flip-flop Q output would appear if it is low. Therefore, it is possible to implement Mealy sequential systems and to increment the number of variables of a product over the number of inputs of an AND gate.

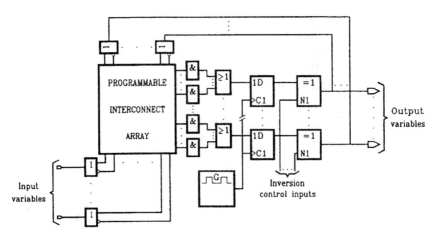

FIGURE 2.21 *Basic PAL-based PLS with programmable output polarity by means of an exclusive OR gate.*

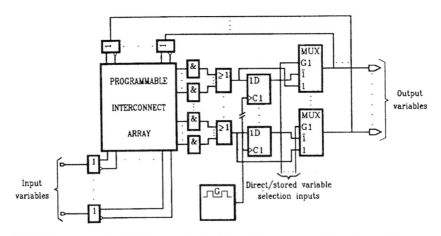

FIGURE 2.22 *Basic PAL-based PLS with a flip-flop and direct/stored OR gate outputs multiplexing.*

Figure 2.23 is a combination of Figures 2.19 to 2.22. All the elements associated with every OR gate output constitute a functional block called a **macrocell** (Figure 2.24).

The macrocell concept leads to the system illustrated in Figure 2.25, that represents the block diagram of the PAL-based programmable logic device (PLD).

However, there are many other implementation alternatives to the macrocell shown in Figure 2.24. One of them appears in Figure 2.26, using an inverter and a multiplexer (MUX2) instead of the exclusive OR gate of Figure 2.24.

PROGRAMMABLE LOGIC DEVICES

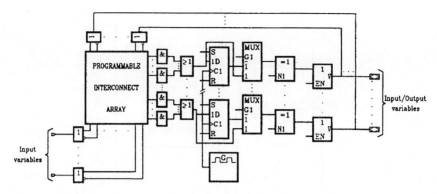

FIGURE 2.23 *PAL-based PLS that combines all the features of Figures 2.19 to 2.22.*

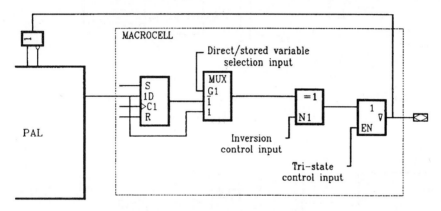

FIGURE 2.24 *Macrocell concept.*

An analysis of the macrocell (Figures 2.24 and 2.26) reveals that the digital signals can be divide into two groups:

1. **Data signals**, which are the outputs of the PAL OR gates (stored and transmitted internally through the macrocell) and input/output (I/O) external pins.
2. **Control signals** that allow the selection of the macrocell operation mode, i.e. the way it stores and transmits data signals. For instance, in Figure 2.26 the control signals are:
    - D-type flip-flop reset (R), set (S) and clock (C1) inputs;
    - multiplexers MUX1 and MUX2 selection inputs (G1);
    - tri-state gate enable signal (EN).

Control signals can be implemented in three different ways:
   - array-independent programmability, by means of an element of selectable conduction/non-conduction states;

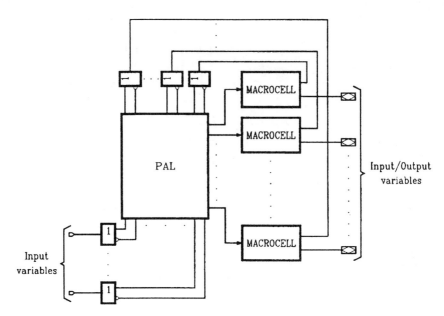

FIGURE 2.25 *Block diagram of a PAL-based programmable logic device.*

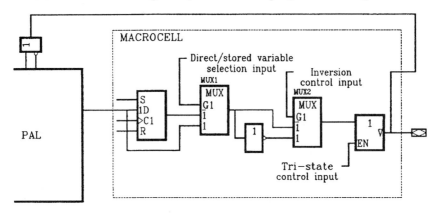

FIGURE 2.26 *Alternative implementation of Figure 2.24 macrocell using a multiplexer (MUX2) instead of an exclusive OR gate.*

- array-dependent programmability. In this case the array is divided into two parts:
  - a PAL, that generates the data signals connected to D-type flip-flop inputs;
  - an AND gate array that generates control signals. The existence of this array means that the PLD block diagram of Figure 2.25 becomes that of Figure 2.27;
- direct connection to an external pin.

PROGRAMMABLE LOGIC DEVICES 65

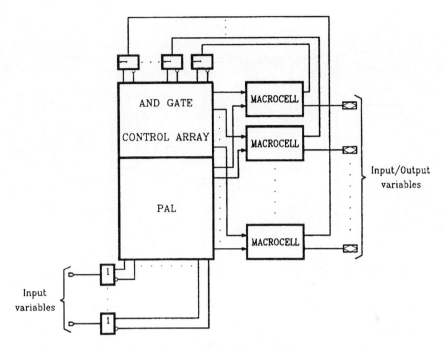

FIGURE 2.27 *Block diagram of a PLD with an AND gate array that generates macrocell control signals.*

The PLD macrocells can combine these three methods of implementation, so there is a variety of possible alternatives. Figure 2.28 shows one where the flip-flop R and S inputs are connected to the AND gate array, the MUX1 and MUX2 multiplexers selection inputs (G1) and the tri-state gate enable signal have independent programmability, and the flip-flop clock input (C) is connected to an external pin.

Figure 2.29 constitutes another alternative where the tri-state gate enable signal is connected to the AND gate array rather than to a programmable element.

On the other hand, in Figures 2.23, 2.24, 2.26, 2.28 and 2.29, the tri-state gate output connected to the I/O pin is fed back directly to the PAL. As a result, a large delay appears when the flip-flop generates an internal state variable because the feedback path includes the multiplexer MUX1, the exclusive OR gate or the multiplexer MUX2, and the tri-state gate.

Figure 2.30 shows a better implementation using a third multiplexer (MUX3) which reduces the flip-flop feedback path thereby decreasing the delay. If MUX3 selection input (G1) is low, flip-flop output $\overline{Q}$ is fed back to the array. When G1 is high, the I/O pin level is sent to the array. As MUX1 and MUX3 selection inputs are connected, when the OR gate

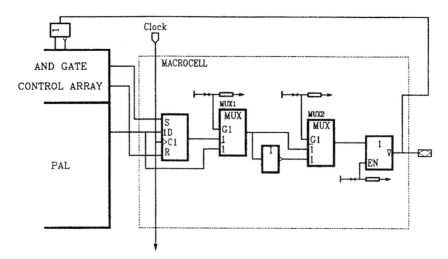

FIGURE 2.28 *Example of macrocell control signals implementation.*

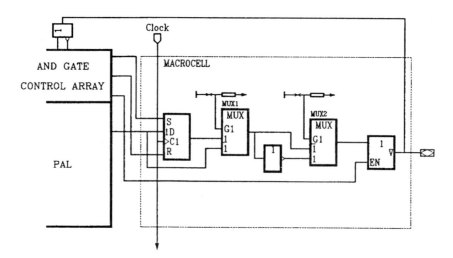

FIGURE 2.29 *Example of macrocell control signals implementation.*

output appears at the MUX1 output, the I/O pin is fed to the array, and when the flip-flop Q output appears at MUX1 output, $\bar{Q}$ is fed back to the array.

Another alternative is shown in Figure 2.31, where MUX1 and MUX3 have independent selection variables, and MUX1 output is connected to MUX3 input 1. Using this structure it is possible to feed back the OR gate output or the flip-flop output. The reader can imagine more alternatives.

PROGRAMMABLE LOGIC DEVICES

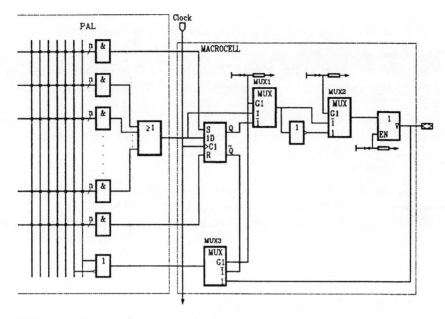

FIGURE 2.30 *Macrocell with a multiplexer (MUX3) to reduce feedback path delay.*

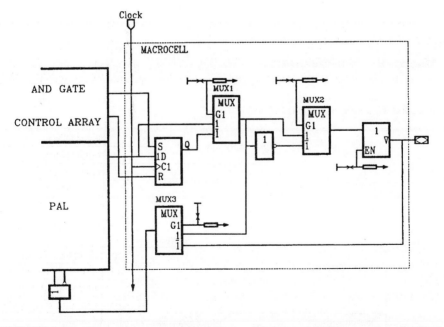

FIGURE 2.31 *Alternative macrocell with a multiplexer (MUX3) to reduce feedback path delay.*

Besides their differences, the macrocells represented in Figures 2.24, 2.26, 2.28, 2.29, 2.30 and 2.31 have, however, some features in common:

1. They have only one flip-flop.
2. They have only one feedback path. Therefore, it is impossible to feed back the flip-flop output at the same time as the I/O pin is being used as an input.
3. The complexity of the logic function assigned to one macrocell is limited (without multiple feedback, i.e. larger delay), the reason being that its input is connected to only one OR gate output.
4. They have only one PIA.

We can conclude that it would be useful to increase macrocell complexity, to improve its flexibility and extend the application field of PLDs.

Because there are so many implementation options, a method has been developed to systematize the study of PLDs, and this is the method we use to define our own analysis below.

2.4 Advanced PAL-based PLDs

We can define advanced PAL-based PLDs as those that combine a set of macrocells with a PIA, divided into two parts: a PAL and a control AND gate array.

As shown in Table 2.3, there are three basic concepts to be analyzed in an advanced PLD:

1. **Macrocell implementation**. The macrocell can have one or more (usually two) feedback paths, and one or more (usually two) flip-flops.

TABLE 2.3 *Classification of advanced PLDs*

PAL based advanced PLDs	MACROCELL IMPLEMENTATION	ONE FEEDBACK PATH		ONE FLIP-FLOP
				TWO FLIP-FLOPS
		TWO FEEDBACK PATHS	INDEPENDENT	ONE FLIP-FLOP
				TWO FLIP-FLOPS
			SHARED	ONE FLIP-FLOP
				TWO FLIP-FLOPS
	ARRAY RESOURCES ALLOCATION	FIXED ALLOCATION		
		VARIABLE ALLOCATION		PRODUCT TERM STEERING
				SUM-OF-PRODUCTS STEERING
				PRODUCT TERM ALLOCATION
				EXPANDER PRODUCT TERM ARRAY
	INTERCONNECT ARRAY IMPLEMENTATION	ONE ARRAY (NON-SEGMENTED PLDS)		
		MULTIPLE ARRAY (SEGMENTED PLDS)		

When there are several feedback paths, these can be either independent or shared between macrocells.

2. **Array resources sharing between macrocells**. A PLD is said to have **fixed allocation** if array resources are not shared between macrocells, and **variable allocation** if the resources are shared between macrocells.

3. **PIA implementation**. The PLD is **non-segmented** if it has only one PIA, and **segmented** if it has two or more.

However, the three groups shown in Table 2.3 are not disjoint, as a PLD can have one or more PIAs, sharing or not sharing the resources of the PAL between macrocells, and have different macrocell implementations.

In the following paragraphs we first study PLDs with fixed allocation and different levels of macrocell complexity. Afterwards, we analyze the different kinds of resources sharing, and finally, we look at segmented PLDs.

2.4.1 Fixed allocation advanced PAL-based PLDs

PLDs with one flip-flop and two feedback paths

The block diagram appears in Figure 2.32. Each macrocell has a double connection to the PIA.

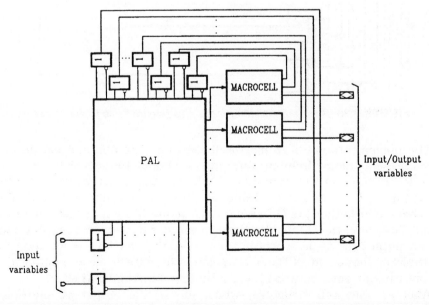

FIGURE 2.32 *Block diagram of a PAL-based PLD with one flip-flop and two feedback paths.*

The basic macrocell is shown in Figure 2.33, where each feedback path uses an independent follower/inverter gate. The input of one of them is connected to the D-type flip-flop Q output, while the input of the other comes from the output of the tri-state gate (also connected to an I/O pin). This structure allows the flip-flop output to be fed back, while at the same time the I/O pin is used as an input.

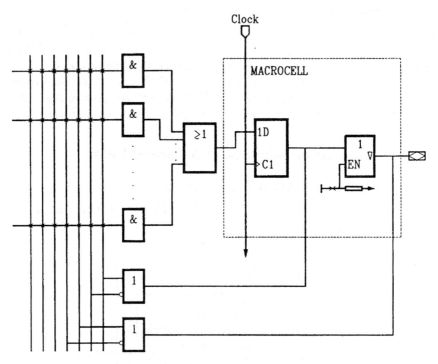

FIGURE 2.33 *Basic macrocell with one flip-flop and two feedback paths.*

The implementation of a macrocell with one flip-flop and two feedback paths presents many different alternatives. One of these is shown in Figure 2.34, where flip-flop and tri-state gate (the one connected to I/O pin) outputs are fed back. Tri-state gates N1 and N2 constitute a multiplexer that allows the choice of whether the OR gate output or the I/O pin appears at flip-flop input. Therefore, the D-type flip-flop can act as a buried state register and/or as an output or as a synchronization flip-flop for the input data. The multiplexer formed by N3 and N4 selects the data to be presented at the output tri-state gate, between D-type flip-flop input and output.

Another macrocell structure, which allows the output of multiplexer N3–N4 to be fed back, is shown in Figure 2.35. Using this structure, it is possible to select whether stored or non-stored signals are fed back.

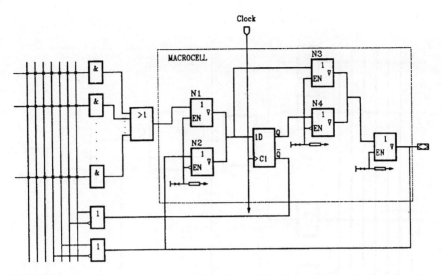

FIGURE 2.34 *Alternative macrocell with one flip-flop and two feedback paths.*

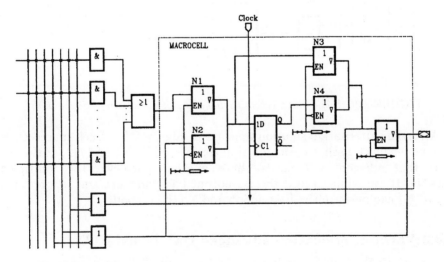

FIGURE 2.35 *Alternative macrocell with one flip-flop and two feedback paths.*

PLDs with two flip-flops and two feedback paths

As can be seen in the macrocells of Figures 2.34 and 2.35, it is impossible to synchronize the input signal while feeding the buried state register back. This problem is avoided by using a new structure with a second flip-flop to synchronize the input's value (Figure 2.36). This macrocell contains two flip-flops and two independent feedback paths.

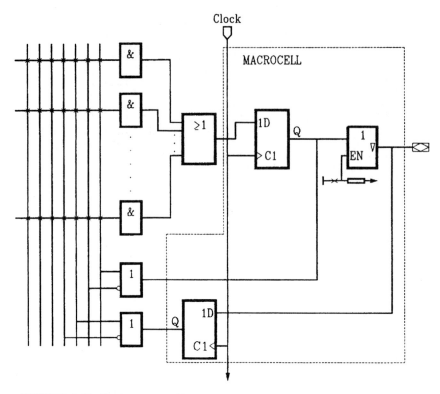

FIGURE 2.36 *Basic macrocell with two flip-flops and two feedback paths.*

A further alternative solution appears in Figure 2.37. Each macrocell has its own feedback path, via the multiplexer formed by N1 and N2, at the same time that adjacent macrocells share another path via the multiplexer N3–N4. If we want to feed the B1 output back and use the I/O pin as an input, N1 can perform the first action and N3 the second.

2.4.2 Variable allocation advanced PAL-based PLDs

These are PLDs that do not have a rigid connection between a macrocell, an OR gate and a certain set of AND gates from the programmable array. This is intended to extend the capabilities of a PAL, where the number of products assigned to each OR gate is fixed.

To achieve the variable allocation it is necessary to share OR gates among macrocells. There are several ways to solve this problem, as we will now see.

First it is necessary to modify the basic block diagram of fixed allocation PLDs (Figure 2.32). This is done by dividing the PAL into several blocks. Figure 2.38 represents a different version of Figure 2.32, dividing the PAL into an AND gate array and a set of OR gates. This is actually the method

PROGRAMMABLE LOGIC DEVICES

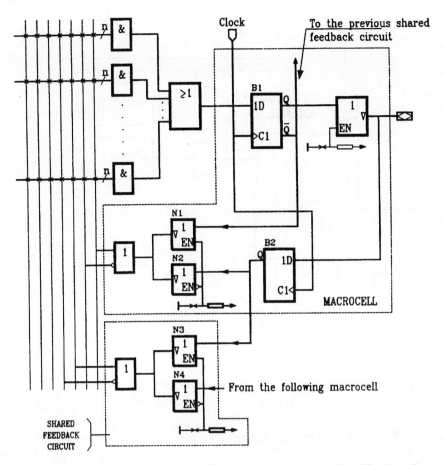

FIGURE 2.37 *Macrocell with two flip-flops and two shared feedback paths.*

adopted by some PLD manufacturers to represent their fixed allocation devices, including the OR gate as a component of the macrocell.

Another version can be found in Figure 2.39, where the PAL is divided into a PIA and several sets of AND gates, each of them connected to an OR gate. By using this block diagram, the different sharing options can be analyzed.

PLDs with logic product steering

This is a solution based on the block diagram of Figure 2.40. More than one AND gate set can be connected to an OR gate, using a steering circuit between them.

This option has a variety of alternatives depending on the number of macrocells that share the same set of AND gates. The solution in Figure 2.41

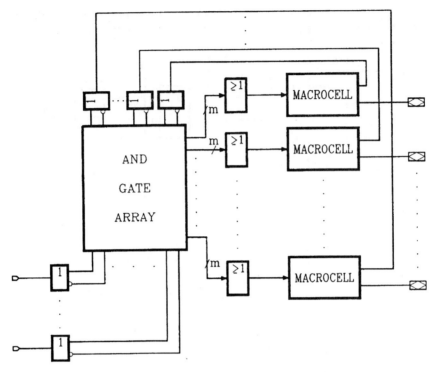

FIGURE 2.38 *Alternative approach of Figure 2.32 PAL-based PLD, dividing the PAL into an AND gate array and a set of OR gates.*

has m AND gates that can be steered between two adjacent macrocells. The steering circuit is an array of m columns (as many as steered AND gates). The OR gates have m inputs, with programmable connections to the columns. The columns have fixed connections to the AND gate outputs. Therefore, this circuit is a small programmable logic array (PLA), that gives the PLD the flexibility it would not have if it were simply implemented with a PAL. Figure 2.42 shows a simplified representation of the steering circuit used by some PLD manufacturers.

PLDs with logic sum-of-products steering

This is another method of variable allocation through sharing OR gates among adjacent macrocells via a multiplexer circuit. The block diagram of this structure is shown in Figure 2.43, and a practical implementation appears in Figure 2.44. The output of each OR gate is connected to one input of a two-channel multiplexer. The other input is connected to a low logic level. The selection input (EN) of the multiplexer (array-independent programmed in Figure 2.44) controls whether the low logic level or the output

PROGRAMMABLE LOGIC DEVICES 75

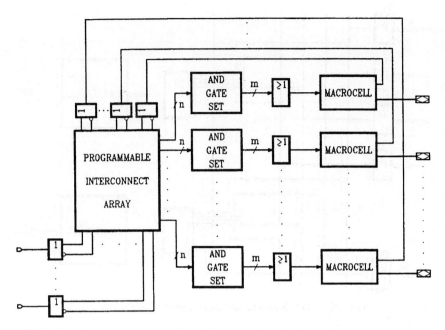

FIGURE 2.39 *Alternative approach of Figure 2.32 PAL-based PLD, dividing the PAL into a PIA and several sets of AND gates, each one connected to an OR gate.*

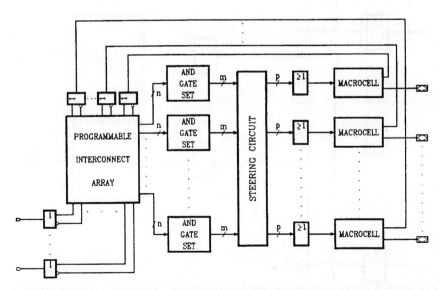

FIGURE 2.40 *Block diagram of PAL-based PLD with logic product steering.*

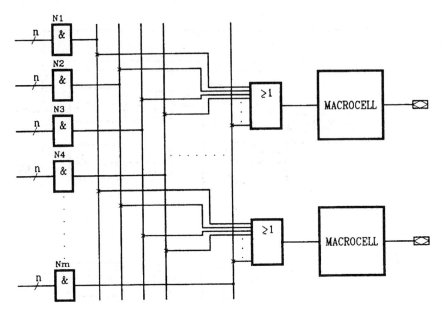

FIGURE 2.41 *Example of logic product steering circuit.*

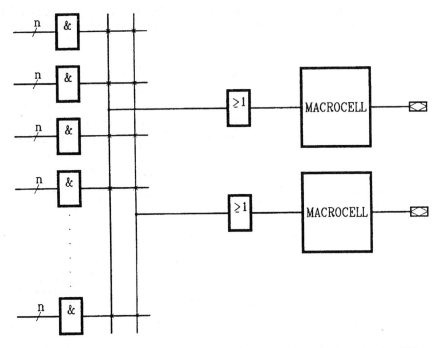

FIGURE 2.42 *Simplified representation of logic product steering circuit of Figure 2.41.*

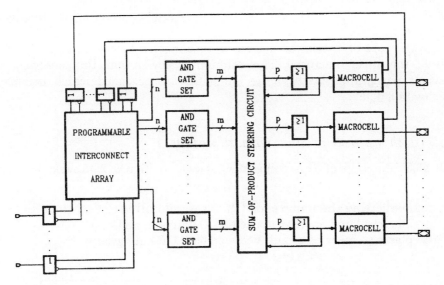

FIGURE 2.43 *Block diagram of a PAL-based PLD with logic sum-of-products steering.*

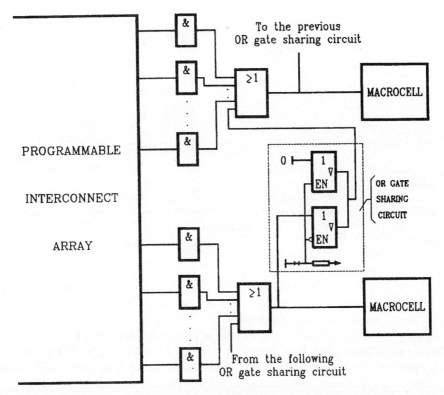

FIGURE 2.44 *Practical implementation of a PAL-based PLD with logic sum-of-products steering.*

of the OR gate of the lower macrocell is connected to the input of the OR gate of the upper macrocell. This solution is slower than that described in previous paragraphs because the OR gates are cascaded which increases propagation delay.

PLDs with multiple product term allocation

The block diagram of this solution is shown in Figure 2.45, and is based on:

- duplicating the number of OR gates, allocating one of them to each macrocell and another to each AND gate set, and
- placing a steering circuit between both OR gate groups.

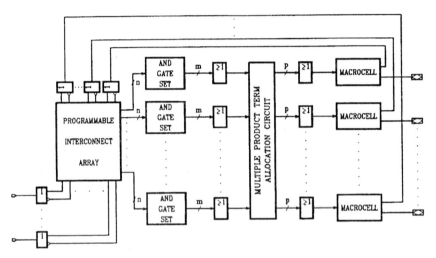

FIGURE 2.45 *Block diagram of a PAL-based PLD with multiple product term allocation.*

It is also possible to use a number of AND gate sets greater than the number of macrocells in such a way that a variable number of AND gates can be assigned to each macrocell. This allows the PLD to be adapted to the conditions of each specific application.

Figure 2.46 represents a practical implementation. Each macrocell is connected to a four-input OR gate. The number of OR gates placed on the left of the product term allocation circuit is twice the number of macrocells. Each of these OR gates has an associated demultiplexer in the steering circuit. The selection input of this demultiplexer (G1) controls whether the output of the OR gate is connected either to a specific macrocell or to the adjacent

PROGRAMMABLE LOGIC DEVICES

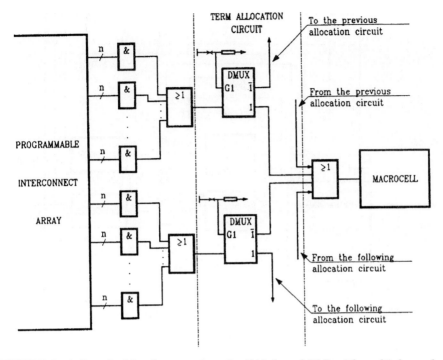

FIGURE 2.46 *Practical implementation of a PAL-based PLD with multiple product term allocation.*

macrocell. In this way we can distribute the resources of the PIA to the appropriate macrocell.

PLDs with expander product term array

As can be seen in the block diagram of Figure 2.47, this solution incorporates the use of NAND gates placed as feedback paths to the PIA.

Another representation can be made, allocating a set of NAND gates to each macrocell, as shown in Figure 2.48. The PLD can be considered as a group of n functional blocks. Each of these groups comprises a set of AND gates, a set of expander NAND gates, one OR gate and one macrocell.

Expander NAND gates have the following advantages:

1. It is possible to increase the complexity of the logic functions associated with a macrocell.
2. The flexibility of the device is increased because the gate is able to implement different combinational or sequential circuits (input synchronization registers, level- or edge-triggered, etc.).

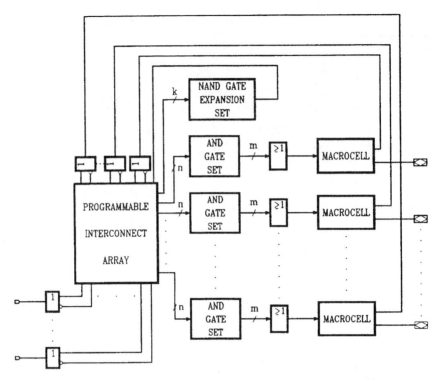

FIGURE 2.47 *Block diagram of a PAL-based PLD with expander product term array.*

2.4.3 PLDs with multiple arrays

The common structure of all previously mentioned PLDs appears in Figure 2.49, where each macrocell has its feedback path connected to a unique array. In this structure, when the number of inputs and macrocells in a PLD increases, the number of AND gate inputs does the same, and so too does the circuit propagation delay.

On the other hand, in many applications, when there is a high number of macrocells, it is not necessary to feed the state of each macrocell back to all the others. These are the main reasons for the development of segmented PLDs. One example is shown in Figure 2.50.

In these devices we can distinguish two kinds of array: a global one (global bus) and several local ones (local buses). The global array is connected to all the input variables and to the output variable of a limited number of macrocells. Each local array is connected to a group of macrocells (that are not connected to the global array), and therefore there are only local feedback paths for each group.

The use of segmentation in addition to the other concepts discussed makes

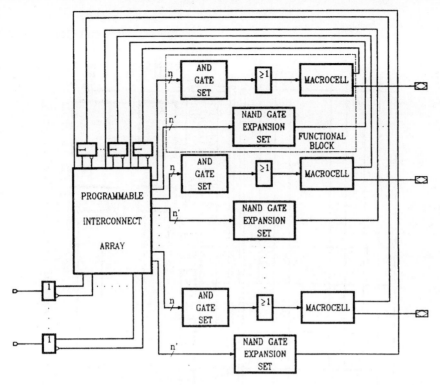

FIGURE 2.48 *Functional block diagram of a PAL-based PLD including expansion NAND gates.*

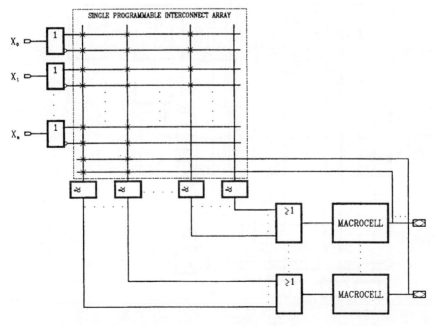

FIGURE 2.49 *Basic structure of a non-segmented PAL-based PLD.*

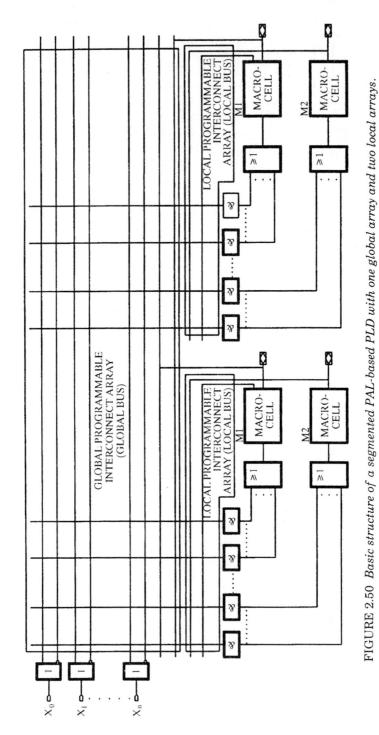

FIGURE 2.50 *Basic structure of a segmented PAL-based PLD with one global array and two local arrays.*

PROGRAMMABLE LOGIC DEVICES

available a large number of alternative PLD implementations. For instance, as in non-segmented structures, macrocells can have one or more feedback paths. A segmented structure with two feedback paths allows macrocells to be fed back through either global or local buses, and I/O pins to be used as inputs to the global bus (PLD inputs), while the state of the macrocell is fed back through the local bus. This is the solution presented in Figure 2.51.

The many possible implementations of advanced PLDs explain the different names that manufacturers give to their devices. A sample of names for interconnect arrays from different manufacturers is presented in Table 2.4.

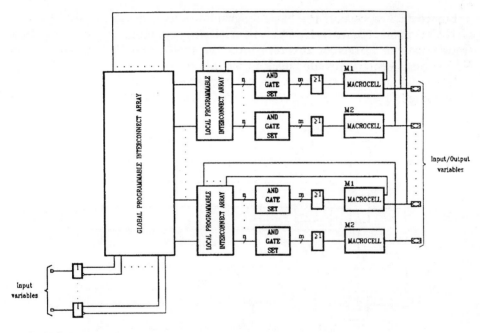

FIGURE 2.51 *Example of PAL-based PLD with one global array and two local arrays.*

TABLE 2.4 *Some commercial references for connection arrays*

ARCHITECTURE	COMPANY	REFERENCE
NON-SEGMENTED	ALTERA	PROGRAMMABLE INTERCONNECT ARRAY (PIA)
	AMD	SWITCH MATRIX
	INTEL	BUS
	XILINX	UNIVERSAL INTERCONNECT MATRIX (UIM)
SEGMENTED	INTEL	GLOBAL BUS / LOCAL BUS
	ATMEL	TOTAL BUS / REGIONAL BUS

2.5 PLDs using universal gates

In section 2.2 several programmable logic arrays implemented with AND and OR gates were studied. These arrays allow the implementation of any combinational system. Besides feeding back the outputs to the inputs, directly or through a synchronous register, they are converted into sequential programmable logic devices, as studied in section 2.3.

In this section, we are going to analyze programmable arrays using universal gates which allow the implementation of digital systems, either combinational or sequential, with only one array. There is no universally accepted name for these circuits and they vary among manufacturers [PHIL 94].

Figure 2.52 represents the basic diagram of such a device, formed by a set of NAND gates, characterized by the existence of a single array whose lines (rows) are the (direct or inverse) input variables and the outputs of the NAND gates which are folded back to the inputs.

By elimination of the appropriate connections, any logic function can be obtained. Figure 2.53 represents a foldback NAND gate array which

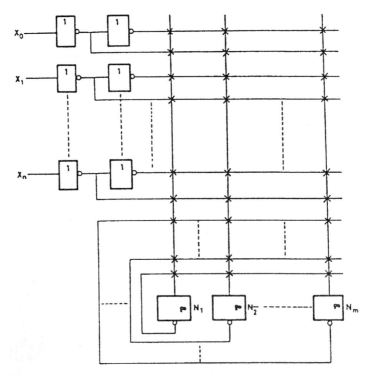

FIGURE 2.52 *Programmable logic array using foldback NAND gates.*

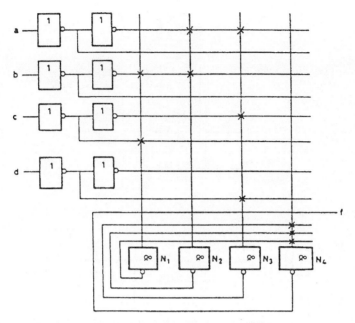

FIGURE 2.53 *Implementation of function $f = b\bar{c} + ab + ac\bar{d}$ using a foldback NAND gate PLA.*

implements the function $f = b\bar{c} + ab + ac\bar{d}$, which is equivalent to:

$$f = \overline{\overline{b\bar{c}}\,\overline{ab}\,\overline{ac\bar{d}}}$$

For this purpose, the array must have a minimum of four NAND gates, one for each of the three products and the fourth to generate function f.

In Figure 2.53 gates N_1 to N_3 generate the three inverted products and N_4 generates function f.

Similarly, it is possible to implement any sequential system. For example the rising edge-triggered D flip-flop of Figure 2.54a is implemented with a feedback NAND gate array as shown in Figure 2.54b.

In Figure 2.53, we see that the output of N_4 is not connected to any input gate once programming is completed. On the other hand, it is convenient that the output variables have the maximum possible fan-out. Therefore, a real programmable logic array using feedback NAND gates is as shown in Figure 2.55, where there are two sets of NAND gates: a set of m_1 foldback gates and a set of m_2 non-foldback gates.

PLDs made of NAND gates also allow many variations such as:

1. The inclusion of synchronous flip-flops in all or part of the outputs of the NAND foldback gates to simplify the design of synchronous sequential systems. There are multiple options which differ on the type of flip-flop employed, the programmability of the asynchronous reset inputs, the

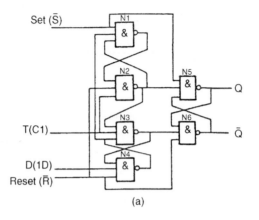
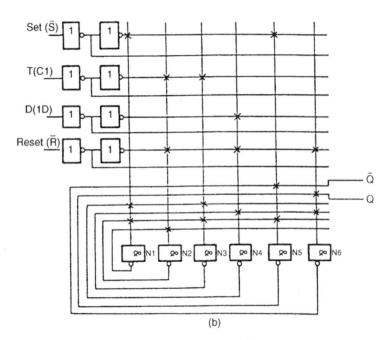

FIGURE 2.54 *Edge-triggered D flip-flop implementation using a foldback NAND gate PLA.* (a) *Edge-triggered D flip-flop circuit diagram.* (b) *Foldback NAND gate PLA implementation.*

programmability of the pulse generator, etc. Figure 2.56 depicts a foldback NAND gate array using:
- m_1 NAND gate outputs directly fold back (N_1 to N_{m1});
- m_2 NAND gate outputs fold back via edge-triggered D flip-flops (N''_1 to N''_{m2});

PROGRAMMABLE LOGIC DEVICES

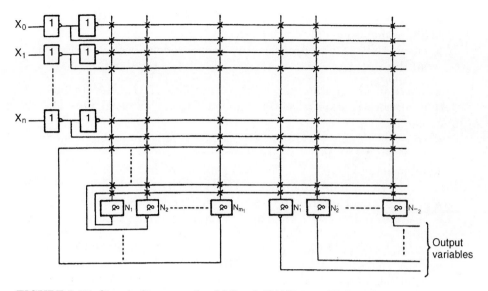

FIGURE 2.55 *Circuit diagram of a foldback NAND gate PLA with an output array.*

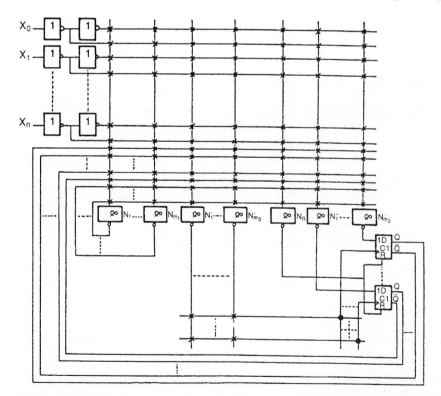

FIGURE 2.56 *Foldback NAND gate PLA with feedback edge-triggered D flip-flops.*

- m_3 NAND gates forming a second array with the C inputs of the above-mentioned D flip-flops (N'_1 to N'_{m3});
- one NAND gate to program the asynchronous reset (R), common to all flip-flops.

2. The insertion of tri-state gates at the outputs of the NAND gates. The output of the tri-state gates is connected to an external pin. The third state can be:
    - constant during operation (e.g. set by a programmable element as a fuse or a MOS floating gate transistor);
    - programmable (e.g. via a NAND gate of the array).

Figure 2.57 represents a foldback NAND gate array having:
- m_1 directly or internally foldback NAND gates (N_1 to N_{m1});
- m_2 NAND gates with the outputs connected to an external pin via tri-state gates N'_1 to N'_{m2};
- m_2 NAND gates connected to the tri-state enable inputs of the m_2 gates (N''_1 to N''_{m2});
- foldback of each of the m_2 output pins to the array via pairs of inverters.

The output pins Y_1 to Y_{m2} whose gate is held constant at the third state, may be used as input pins. That is why the circuit of Figure 2.57 has $n+1$ input pins (X_0 to X_n) and m_2 pins (Y_1 to Y_{m2}) that can be inputs or outputs.

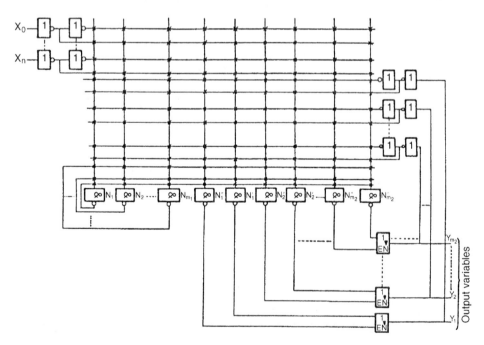

FIGURE 2.57 *Foldback NAND gate PLA with input/output pins.*

3. The insertion of synchronous flip-flops between the output of the NAND gates and the output pins. In the array of Figure 2.57 these flip-flops can be placed between the gate outputs and the foldback connections, and they may form an input register (to synchronize the corresponding input variables) or an output register. Using these flip-flops we simplify the design of programmable logic controllers.

Programmable logic arrays with foldback NAND gates have the following advantages over AND–OR programmable logic arrays:
- shorter propagation time because two NAND gates have a smaller delay than an AND gate followed by an OR gate;
- smaller number of pins to implement a digital system because of the capability of internal foldback.

2.6 PLD technologies

PLDs are manufactured in the transistor–transistor low power Schottky technology (LSTTL) and in the complementary MOS transistors (CMOS) technology.

In LSTTL technology the programmable elements consist of high precision fuses. One-time programming is achieved by blowing the fuse by means of the injection of a controlled current with a value higher than the normal operating one. In this technology, we achieve high speed, one-time programmable synchronous logic controllers with clock frequencies up to 60 MHz.

In the CMOS technology, the programmable element is a floating gate transistor and programming is achieved by injecting charges in the floating gate by means of an applied voltage higher than the normal operation value. Just like reprogrammable read-only memories (RPROM) [MAND 91] [WAKE 94], there are two versions, one erasable by ultraviolet light and the other by electrical pulses.

These PLDs are reprogrammable a great number of times. They have high-level noise immunity, low power consumption and high integration density. They also allow the implementation of synchronous logic controllers with clock frequencies up to 66 MHz [INTE 91].

Bibliography

[ALMA 94]   A.E.A. Almaini, *Electronic Logic Systems*, 3rd edition. Prentice Hall, 1994.
[FLOY 94]   T.L. Floyd, *Digital Fundamentals*, 5th edition. Prentice Hall, 1994.
[INTE 91]   Intel Corporation, *Programmable Logic*, 1991.
[MAND 91]   E. Mandado, *Sistemas electrónicos digitales*, 7th edition. Marcombo, 1991.
[McCL 86]   J.McCluskey, *Logic Design Principles*. Prentice Hall, 1986.

[PHIL 87]   Philips, *Semi-custom Programmable Logic Devices*, IC3, 1987.
[PHIL 94]   Philips, *Programmable Logic Devices*, IC13 data handbook, 1994.
[PROS87]   F.P. Prosser and D.E. Ninkel, *The Art of Digital Design*. Prentice Hall, 1987.
[TEXA 89]   Texas Instruments, *The TTL Data Book*, volume 2, *Advanced Low-power Schottky*, 1989.
[WAKE 94]   J.F. Wakerly, *Digital Design*, 2nd edition. Prentice Hall, 1994.

CHAPTER 3

Logic controller design using programmable logic devices

3.1 Introduction

In the previous chapter, different programmable logic devices (PLDs) and their characteristics were studied. The design of several small digital systems was also described. Nevertheless, the application of PLDs to the design of complex systems is quite difficult (if not impossible) without using computer aided design (CAD) methods, that provide:

- cost reduction and systematic design;
- increased reliability of the whole system;
- easy implementation of very complex digital systems.

3.2 Workstation for logic controller design using PLDs

A design workstation consists of the following components:

1. A general-purpose computer (compatible personal computer, minicomputer, etc.).
2. Design software.
3. Programming hardware.

This hardware can be implemented in two different ways:

1. A printed circuit board connected to the computer bus and an external coupling module (Figure 3.1).
2. An independent peripheral system connected to the computer by a standard serial connection (usually RS-232C) (Figure 3.2).

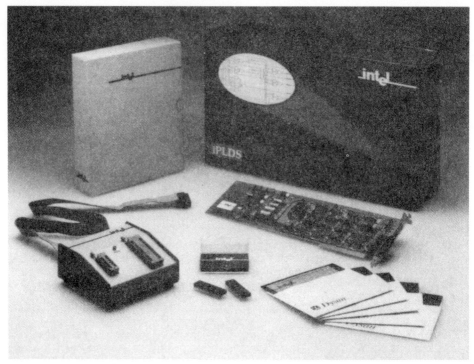

FIGURE 3.1 *Logic controller CAD system using a PC with an internal programming board (courtesy of Intel).*

3.3 Logic controller design phases using PLDs

The phases of computer aided design of logic controllers using PLDs are shown in Table 3.1. These phases are studied in the paragraphs that follow.

3.3.1 Circuit capture

The circuit capture phase consists of the graphic representation of the circuit specifications in such a way that they can be converted to logic equations.

There are several approaches to circuit capture:

1. By means of a **logic diagram**. Sometimes it is necessary to implement using PLDs a logic controller that has been previously designed with SSI and/or MSI integrated circuits. In these cases it is possible to draw the logic diagram on screen by means of graphic design software (for instance Futurenet, Orcad). These programs convert the circuit specification into a file that is converted to logic equations. This circuit capture method can be used for both combinational and sequential logic controllers.

FIGURE 3.2 *Logic controller CAD system using a PC and an externally connected commercial device programmer (courtesy of Signetics).*

TABLE 3.1 *Computer aided design of logic controllers using PLDs*

- Circuit description.
- Translation to algebraic equations.
- Equation minimization.
- PLD selection.
- Circuit behaviour simulation.
- Programming file generation.

2. By means of **logic equations**. This approach could be used for both combinational and sequential systems but it is mostly used for the combinational only. Generally the combinational system is specified by means of a truth table from which a sum-of-products (or a product-of-sums) is obtained. These expressions are stored in a file which is then processed by a minimization program.

   This capture method can only be used for sequential circuits if they are already designed and their logic equations are available. If only the system specifications are available to the designer, this method has no advantages because it is then necessary to generate the logic equations by hand.

3. By means of a **state diagram**. This approach is only valid for sequential circuits because combinational circuits do not have any internal states. This method constitutes a powerful CAD tool, especially when the number of input variables and sequences is large.

   The behaviour of the sequential systems is specified by means of a state diagram directly obtained from an operation flow diagram. The state transitions are defined by means of high-level language statements. Figure 3.3 shows an example. The file containing these statements is then translated to logic equations by means of a conversion program.

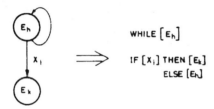

FIGURE 3.3 *Internal state change conditions specified in a high-level design language.*

3.3.2 Conversion to logic equations

The conversion phase is only needed if the circuit has been specified by means of its logic diagram or its state diagram (sequential systems only). A translation program does the conversion and generates a file containing the logic equations to be used in the next phase.

3.3.3 Logic equation minimization

The minimization phase is implemented by means of a program using the file that contains the logic equations (generated in the previous phase). It obtains a minimal sum-of-products equation to be programmed into a PLA or a PAL.

3.3.4 PLD selection

The PLD selection is done by choosing a device with the appropriate capability to handle the required number of input, output and internal state variables.

3.3.5 Circuit behaviour simulation

Circuit behaviour simulation is achieved by applying input combinations in a determined sequence and verifying the operation of the simulated circuit by observation of the internal states and output variables. Typically, the simulation programs have two operation modes:

1. Entering of the input combinations step by step. This mode is useful for simple circuits but not for complex ones because of the large number of input combination sequences that it is necessary to apply.
2. Automatic generation and application of input combinations (by the simulation program) that covers at least 95 per cent of possible faults.

3.3.6 Programming file generation

When the simulation is finished, the minimum logic equations are converted into a table that specifies the programmable elements that must be activated. The programming module uses this table to program the device.

3.4 Logic controller design using PLDs

To understand the methods described in the previous paragraphs some examples are explained below.

3.4.1 Combinational logic controller design using PLDs

Any combinational circuit can be implemented using PLDs. Implementation with random access read-only memory (ROM, PROM, EPROM, EEPROM or FLASH) needs only the first CAD phase previously described (the circuit edition by means of the truth table). Implementation with PLDs, however, uses most of the CAD phases. Let us look at an example.

EXAMPLE 3.1

The fuel-oil tank T of Figure 3.4 is at a certain service temperature by means of an electric heater E. A pump P sends the fuel-oil into two gas burners (B1 and B2) of a furnace.

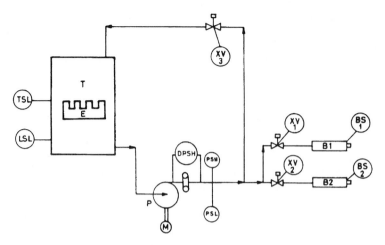

FIGURE 3.4 *Fuel-oil tank*.

It is necessary to implement automatic supervision of the process by means of a combinational circuit in accordance with the following set of specifications:

1. If the tank level goes below a certain value (LSL = 1), the pump must stop (PP = 1), the electrovalve XV3 must open (XV3 = 1) and the red lamp RL must be activated to indicate that the system is out of service (RL = 1).
2. If the temperature of the tank goes below a set value (TSL = 1), the same actions as in (1) above must take place.
3. If the pressure loss in filter F increases over a certain value (DPSH = 1) or the pressure in the fuel-oil collector goes under a certain value (PSL = 1), the same actions as in (1) above must take place.
4. If the pressure in the fuel-oil collector goes over a certain value (PSH = 1), the electrovalve XV3 must be opened (XV3 = 1).
5. If any burner has no flame (BS1 = 1 or BS2 = 1), the corresponding electrovalve XV1 or XV2 must be closed (XV1 = 0 or XV2 = 0) and electrovalve XV3 must be opened (XV3 = 1).
6. If no burner has a flame (BS1 = 1 and BS2 = 1), the system must be set out of service by performing the same actions as in (1) above.
7. In normal operation conditions a green lamp GL must be activated (GL = 1).

Solution
From the specifications we generate the truth table shown in Table 3.2.

1. **Implementation with random access memory**. This memory should have a minimum capacity of $2^7 = 128$ addresses of 6 bits. Figure 3.5 shows the standard graphic symbol of a PROM memory. The input variables are connected to the address inputs and the output variables are the data outputs of the memory. From

LOGIC CONTROLLER DESIGN USING PLDS

TABLE 3.2

	LSL	TSL	DPSH	PSL	PSH	BS1	BS2	PP	XV1	XV2	XV3	GL	RL
A	0	0	0	0	0	0	0	0	1	1	0	1	0
B	1	X	X	X	X	0	0	1	1	1	1	0	1
C	1	X	X	X	X	1	0	1	0	1	1	0	1
D	1	X	X	X	X	0	1	1	1	0	1	0	1
E	1	X	X	X	X	1	1	1	0	0	1	0	1
F	X	1	X	X	X	0	0	1	1	1	1	0	1
G	X	1	X	X	X	1	0	1	0	1	1	0	1
H	X	1	X	X	X	0	1	1	1	0	1	0	1
I	X	1	X	X	X	1	1	1	0	0	1	0	1
J	X	X	1	X	X	0	0	1	1	1	1	0	1
K	X	X	1	X	X	1	0	1	0	1	1	0	1
L	X	X	1	X	X	0	1	1	1	0	1	0	1
M	X	X	1	X	X	1	1	1	0	0	1	0	1
N	X	X	X	1	X	0	0	1	1	1	1	0	1
O	X	X	X	1	X	1	0	1	0	1	1	0	1
P	X	X	X	1	X	0	1	1	1	0	1	0	1
Q	X	X	X	1	X	1	1	1	0	0	1	0	1
R	0	0	0	0	1	0	0	0	1	1	1	1	0
S	0	0	0	0	X	1	0	0	0	1	1	1	0
T	0	0	0	0	X	0	1	0	1	0	1	1	0
U	X	X	X	X	X	1	1	1	0	0	1	0	1

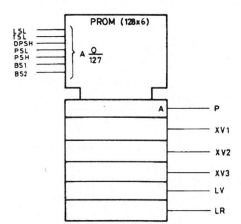

FIGURE 3.5 *PROM memory implementation of the logic controller for the fuel-oil tank of Figure 3.4.*

TABLE 3.3

A_6	A_5	A_4	A_3	A_2	A_1	A_0	D_5	D_4	D_3	D_2	D_1	D_0
0	0	0	0	0	0	0	0	1	1	0	1	0
0	0	0	0	0	0	1	0	1	0	1	1	0
0	0	0	0	0	1	0	0	0	1	1	1	0
0	0	0	0	0	1	1	1	0	0	1	0	1
0	0	0	0	1	0	0	0	1	1	1	1	0
0	0	0	0	1	0	1	0	1	0	1	1	0
0	0	0	0	1	1	0	0	0	1	1	1	0
0	0	0	0	1	1	1	1	0	0	1	0	1
0	0	0	1	0	0	0	1	1	1	1	0	1
0	0	0	1	0	0	1	1	1	0	1	0	1
0	0	0	1	0	1	0	1	0	1	1	0	1
0	0	0	1	0	1	1	1	0	0	1	0	1
0	0	0	1	1	0	0	1	1	1	1	0	1
0	0	0	1	1	0	1	1	1	0	1	0	1
0	0	0	1	1	1	0	1	0	1	1	0	1
0	0	0	1	1	1	1	1	0	0	1	0	1
0	0	1	0	0	0	0	1	1	1	1	0	1
0	0	1	0	0	0	1	1	1	0	1	0	1
0	0	1	0	0	1	0	1	0	1	1	0	1
0	0	1	0	0	1	1	1	0	0	1	0	1
0	0	1	0	1	0	0	1	1	1	1	0	1
0	0	1	0	1	0	1	1	1	0	1	0	1
0	0	1	0	1	1	0	1	0	1	1	0	1
0	0	1	0	1	1	1	1	0	0	1	0	1
0	0	1	1	0	0	0	1	1	1	1	0	1
0	0	1	1	0	0	1	1	1	0	1	0	1
0	0	1	1	0	1	0	1	0	1	1	0	1
0	0	1	1	0	1	1	1	0	0	1	0	1
0	0	1	1	1	0	0	1	1	1	1	0	1
0	0	1	1	1	0	1	1	1	0	1	0	1
0	0	1	1	1	1	0	1	0	1	1	0	1
0	0	1	1	1	1	1	1	0	0	1	0	1
0	1	0	0	0	0	0	1	1	1	1	0	1
0	1	0	0	0	0	1	1	1	0	1	0	1
0	1	0	0	0	1	0	1	0	1	1	0	1
0	1	0	0	0	1	1	1	0	0	1	0	1
0	1	0	0	1	0	0	1	1	1	1	0	1
0	1	0	0	1	0	1	1	1	0	1	0	1
0	1	0	0	1	1	0	1	0	1	1	0	1
0	1	0	0	1	1	1	1	0	0	1	0	1
0	1	0	1	0	0	0	1	1	1	1	0	1
0	1	0	1	0	0	1	1	1	0	1	0	1
0	1	0	1	0	1	0	1	0	1	1	0	1
0	1	0	1	0	1	1	1	0	0	1	0	1
0	1	0	1	1	0	0	1	1	1	1	0	1
0	1	0	1	1	0	1	1	1	0	1	0	1
0	1	0	1	1	1	0	1	0	1	1	0	1
0	1	0	1	1	1	1	1	0	0	1	0	1
0	1	1	0	0	0	0	1	1	1	1	0	1
0	1	1	0	0	0	1	1	1	0	1	0	1
0	1	1	0	0	1	0	1	0	1	1	0	1
0	1	1	0	0	1	1	1	0	0	1	0	1
0	1	1	0	1	0	0	1	1	1	1	0	1
0	1	1	0	1	0	1	1	1	0	1	0	1
0	1	1	0	1	1	0	1	0	1	1	0	1
0	1	1	0	1	1	1	1	0	0	1	0	1
0	1	1	1	0	0	0	1	1	1	1	0	1
0	1	1	1	0	0	1	1	1	0	1	0	1
0	1	1	1	0	1	0	1	0	1	1	0	1
0	1	1	1	0	1	1	1	0	0	1	0	1
0	1	1	1	1	0	0	1	1	1	1	0	1
0	1	1	1	1	0	1	1	1	0	1	0	1
0	1	1	1	1	1	0	1	0	1	1	0	1
0	1	1	1	1	1	1	1	0	0	1	0	1
1	0	0	0	0	0	0	1	1	1	1	0	1
1	0	0	0	0	0	1	1	1	0	1	0	1
1	0	0	0	0	1	0	1	0	1	1	0	1
1	0	0	0	0	1	1	1	0	0	1	0	1
1	0	0	0	1	0	0	1	1	1	1	0	1
1	0	0	0	1	0	1	1	1	0	1	0	1
1	0	0	0	1	1	0	1	0	1	1	0	1
1	0	0	0	1	1	1	1	0	0	1	0	1
1	0	0	1	0	0	0	1	1	1	1	0	1
1	0	0	1	0	0	1	1	1	0	1	0	1
1	0	0	1	0	1	0	1	0	1	1	0	1
1	0	0	1	0	1	1	1	0	0	1	0	1
1	0	0	1	1	0	0	1	1	1	1	0	1
1	0	0	1	1	0	1	1	1	0	1	0	1
1	0	0	1	1	1	0	1	0	1	1	0	1
1	0	0	1	1	1	1	1	0	0	1	0	1
1	0	1	0	0	0	0	1	1	1	1	0	1
1	0	1	0	0	0	1	1	1	0	1	0	1
1	0	1	0	0	1	0	1	0	1	1	0	1
1	0	1	0	0	1	1	1	0	0	1	0	1
1	0	1	0	1	0	0	1	1	1	1	0	1
1	0	1	0	1	0	1	1	1	0	1	0	1
1	0	1	0	1	1	0	1	0	1	1	0	1
1	0	1	0	1	1	1	1	0	0	1	0	1
1	0	1	1	0	0	0	1	1	1	1	0	1
1	0	1	1	0	0	1	1	1	0	1	0	1
1	0	1	1	0	1	0	1	0	1	1	0	1
1	0	1	1	0	1	1	1	0	0	1	0	1
1	0	1	1	1	0	0	1	1	1	1	0	1
1	0	1	1	1	0	1	1	1	0	1	0	1
1	0	1	1	1	1	0	1	0	1	1	0	1
1	0	1	1	1	1	1	1	0	0	1	0	1
1	1	0	0	0	0	0	1	1	1	1	0	1
1	1	0	0	0	0	1	1	1	0	1	0	1
1	1	0	0	0	1	0	1	0	1	1	0	1
1	1	0	0	0	1	1	1	0	0	1	0	1
1	1	0	0	1	0	0	1	1	1	1	0	1
1	1	0	0	1	0	1	1	1	0	1	0	1
1	1	0	0	1	1	0	1	0	1	1	0	1
1	1	0	0	1	1	1	1	0	0	1	0	1
1	1	0	1	0	0	0	1	1	1	1	0	1
1	1	0	1	0	0	1	1	1	0	1	0	1
1	1	0	1	0	1	0	1	0	1	1	0	1
1	1	0	1	0	1	1	1	0	0	1	0	1
1	1	0	1	1	0	0	1	1	1	1	0	1
1	1	0	1	1	0	1	1	1	0	1	0	1
1	1	0	1	1	1	0	1	0	1	1	0	1
1	1	0	1	1	1	1	1	0	0	1	0	1
1	1	1	0	0	0	0	1	1	1	1	0	1
1	1	1	0	0	0	1	1	1	0	1	0	1
1	1	1	0	0	1	0	1	0	1	1	0	1
1	1	1	0	0	1	1	1	0	0	1	0	1
1	1	1	0	1	0	0	1	1	1	1	0	1
1	1	1	0	1	0	1	1	1	0	1	0	1
1	1	1	0	1	1	0	1	0	1	1	0	1
1	1	1	0	1	1	1	1	0	0	1	0	1
1	1	1	1	0	0	0	1	1	1	1	0	1
1	1	1	1	0	0	1	1	1	0	1	0	1
1	1	1	1	0	1	0	1	0	1	1	0	1
1	1	1	1	0	1	1	1	0	0	1	0	1
1	1	1	1	1	0	0	1	1	1	1	0	1
1	1	1	1	1	0	1	1	1	0	1	0	1
1	1	1	1	1	1	0	1	0	1	1	0	1
1	1	1	1	1	1	1	1	0	0	1	0	1

is assignment and the truth table of Table 3.2 we obtain the contents of the memory (see Table 3.3).

- **Implementation with PLDs**. We begin the design by establishing the sum-of-product equations (not minimized). This can be done in two different ways:
  - obtaining the logic product that corresponds to each line of Table 3.2 (a letter is assigned to each line). For example, line B equals the logic product B = LSL BS1 BS2. From these products we obtain the output variable equations, for example GL = A + R + S + T;
  - obtaining, from the specifications, the conditions of input variables that activate every output. For example, from the specifications we establish that the pump output variable PP must be activated if the tank level drops below a certain value (LSL = 1), or the tank temperature drops below a set value (TSL = 1), or the pressure loss in filter F is greater than a certain value (DPSH = 1), or the pressure in the fuel-oil collector goes below a certain value (PLS = 1) and, finally, if the detectors BS1 and BS2 indicate simultaneously that the two burners have no flame. The final expression is:

$$PP = LSL + TSL + DPSH + PSL + BS1 \cdot BS2$$

These logic equations can be minimized using a CAD tool. In the example below, Intel and Signetics design systems are used.

The design with the Intel system is implemented using the 5C031 device (see appendix 3). Table 3.4 shows the design description including input and output variable declarations and the equations directly obtained from the specifications. This description is made by means of any text editor, according to the rules of the IPLSII design package (Intel Programmable Logic Software II). Note that IPLSII indicates the inverted variables using a prime symbol ('). For example inverted LSL is LSL'.

The input pins and the macrocells of the 5C031 device can be defined in different ways (see Table A3.1 in appendix 3).

Table 3.4 describes the controller. The selection of inputs and outputs is made in the NETWORK part. The variables that are directly connected to the inputs are called INP and their graphic symbols are shown in Figure 3.6a. Each output variable is assigned to a macrocell defined as a combinational output without feedback [combinational output, no feedback (CONF)] and their graphic symbols are shown in Figure 3.6b. A complete description of IPLSII is available in [INTE 88].

The logic equations of the logic controller are indicated in the EQUATIONS part of Table 3.4. From these equations the program generates the minimal expressions of Table 3.5. It also assigns the input and output variables to the 5C031 pins. This assignment is shown in Figure 3.7. The program also generates a file that controls the PLD programmer in order to program the 5C031.

TABLE 3.4

Department of Electronic Technology
University of Vigo (Spain)
Combinational Logic Controller of a Fuel-oil Tank

PART:5C031
INPUTS:LSL,TSL,DPSH,PSL,PSH,BS1,BS2
OUTPUTS:PP,XV1,XV2,XV3,GL,RL

NETWORK:
LSL=INP(LSL)
TSL=INP(TSL)
DPSH=INP(DPSH)
PSL=INP(PSL)
PSH=INP(PSH)
BS1=INP(BS1)
BS2=INP(BS2)
PP=CONF(PPc,VCC)
XV1=CONF(XV1c,VCC)
XV2=CONF(XV2c,VCC)
XV3=CONF(XV3c,VCC)
GL=CONF(GLc,VCC)
RL=CONF(RLc,VCC)

EQUATIONS:
PPc=LSL+TSL+DPSH+PSL+BS1*BS2;
XV1c=BS1*BS2;
XV2c=BS1'*BS2;
XV3c=LSL'*TSL'*DPSH'*PSL'*PSH'*BS1'*BS2';
GLc=LSL'*TSL'*DPSH'*PSL'*BS1'*BS2'+LSL'*TSL'*DPSH'*PSL'*BS1*BS2'+
LSL'*TSL'*DPSH'*PSL'*BS1'*BS2;
RLc=LSL+TSL+DPSH+PSL+BS1*BS2;

END$

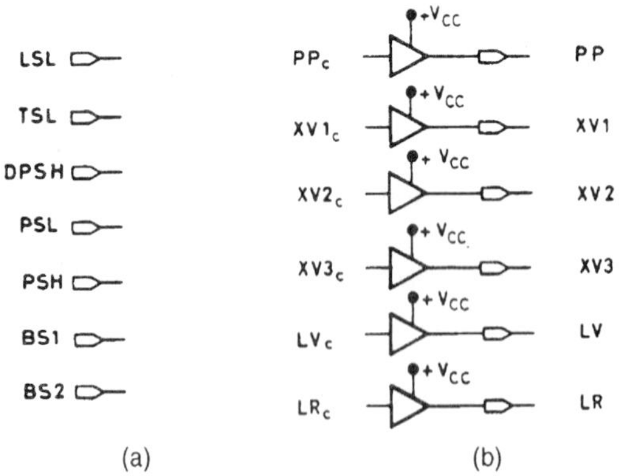

FIGURE 3.6 *IPLSII symbols of inputs and outputs.* (a) *Direct inputs.* (b) *Combinational output, no feedback.*

TABLE 3.5

Department of Electronic Technology
University of Vigo (Spain)
Combinational Logic Controller of a Fuel-oil Tank


```
LEF Version 1.5 Baseline 3.3i
PART:
        5C031
INPUTS:
        LSL, TSL, DPSH, PSL, PSH, BS1, BS2
OUTPUTS:
        PP, XV1, XV2, XV3, GL, RL
NETWORK:
        LSL  = INP(LSL)
        TSL  = INP(TSL)
        DPSH = INP(DPSH)
        PSL  = INP(PSL)
        PSH  = INP(PSH)
        BS1  = INP(BS1)
        BS2  = INP(BS2)
        PP   = CONF(PPc, VCC)
        XV1  = CONF(XV1c, VCC)
        XV2  = CONF(XV2c, VCC)
        XV3  = CONF(XV3c, VCC)
        GL   = CONF(GLc, VCC)
        RL   = CONF(RLc, VCC)
EQUATIONS:
        RLc = (LSL' * TSL' * DPSH' * PSL' * BS2'
            +  LSL' * TSL' * DPSH' * PSL' * BS1')';

        GLc = LSL' * TSL' * DPSH' * PSL' * BS2'
            + LSL' * TSL' * DPSH' * PSL' * BS1';

        XV3c = LSL' * TSL' * DPSH' * PSL' * PSH' * BS1' * BS2';

        XV2c = BS1' * BS2;

        XV1c = BS1 * BS2;

        PPc = LSL' * TSL' * DPSH' * PSL' * BS2'
            + LSL' * TSL' * DPSH' * PSL' * BS1';

END$
```

The design can also be implemented with the PLS100 device from Signetics using the AMAZE CAD system [SIGN 87]. The AMAZE begins by assigning the input and output variables to the circuit pins using a pin list editing program. Figure 3.8 shows the result. The input pins are labelled I, the outputs O, and /CE is the chip enable. The N/C label specifies that the pin is not connected.

The logic equations of the different lines A to U of Table 3.2 and the output variables are edited using the option Edit Logic Boolean Information of the AMAZE editor. The result of this editing is shown in Table 3.6. Note

```
INTEL Logic Optimizing Compiler Utilization Report            TANK.rpt
FIT Version 1.5 Level 3.0i

***** Design implemented successfully

Department of Electronic Technology
University of Vigo (Spain)
Combinational Logic Controller of a Fuel-oil Tank
```

```
               5C031
               - - - -
        BS2 - | 1    20 | - Vcc
        BS1 - | 2    19 | - Gnd
        PSH - | 3    18 | - Gnd
        PSL - | 4    17 | - PP
       DPSH - | 5    16 | - XV1
        TSL - | 6    15 | - XV2
        LSL - | 7    14 | - XV3
        Gnd - | 8    13 | - LV
        Gnd - | 9    12 | - LR
        GND - |10    11 | - Gnd
               - - - -
```

FIGURE 3.7 *PLD 5C031 pin assignment to the input and output variables of the logic controller for the fuel-oil tank.*

```
File Name : tank
Date :
Time :

#################### P I N    L I S T  ####################
     LABEL    ** FNC **PIN - - - - - PIN** FNC **    LABEL
     N/C      ** FE  **  1-|         -28 ** +5V  **VCC
     LSL      ** I   **  2-|         -27 ** I    **N/C
     TSL      ** I   **  3-|         -26 ** I    **N/C
     DPSH     ** I   **  4-|         -25 ** I    **N/C
     PSL      ** I   **  5-|  P      -24 ** I    **N/C
     PSH      ** I   **  6-|  L      -23 ** I    **N/C
     BS1      ** I   **  7-|  S      -22 ** I    **N/C
     BS2      ** I   **  8-|  1      -21 ** I    **N/C
     N/C      ** I   **  9-|  0      -20 ** I    **N/C
     PP       ** O   ** 10-|  0      -19 ** /CE  **N/C
     XV1      ** O   ** 11-|         -18 ** O    **N/C
     XV2      ** O   ** 12-|         -17 ** O    **N/C
     XV3      ** O   ** 13-|         -16 ** O    **RL
     GND      ** 0V  ** 14-|         -15 ** O    **GL
                              - - - -
```

FIGURE 3.8 *PLD PLS100 pin assignments to input and output variables of the logic controller for the fuel-oil tank.*

LOGIC CONTROLLER DESIGN USING PLDS

TABLE 3.6

@DEVICE TYPE
PLS100

@DRAWING

@REVISION

@DATE

@SYMBOL

@COMPANY
Department of Electronic Technology
University of Vigo (Spain)

@NAME
Combinational Logic Controller of a Fuel-oil Tank

@DESCRIPTION

@COMMON PRODUCT TERM
A=/LSL*/TSL*/DPSH*/PSL*/PSH*/BS1*/BS2;
B=LSL*/BS1*/BS2;
C=LSL*BS1*/BS2;
D=LSL*/BS1*BS2;
E=LSL*BS1*BS2;
F=TSL*/BS1*/BS2;
G=TSL*BS1*/BS2;
H=TSL*/BS1*BS2;
I=TSL*BS1*BS2;
J=DPSH*/BS1*/BS2;
K=DPSH*BS1*/BS2;
L=DPSH*/BS1*BS2;
M=DPSH*BS1*BS2;
N=PSL*/BS1*/BS2;
O=PSL*BS1*/BS2;
P=PSL*/BS1*BS2;
Q=PSL*BS1*BS2;
R=/LSL*/TSL*/DPSH*/PSL*PSH*/BS1*/BS2;
S=/LSL*/TSL*/DPSH*/PSL*BS1*/BS2;
T=/LSL*/TSL*/DPSH*/PSL*/BS1*BS2;
U=BS1*BS2;

@LOGIC EQUATION
PP=B+C+D+E+F+G+H+I+J+K+L+M+N+O+P+Q+U;
XV1=A+B+D+F+H+J+L+N+P+R+T;
XV2=A+B+C+F+G+J+K+N+O+R+S;
XV3=/A;
GL=A+R+S+T;
RL=B+C+D+E+F+G+H+I+J+K+L+M+N+O+P+Q+U;

that AMAZE uses the solidus (/) symbol to specify the inverted variable (for example, /LSL).

In the next step, the assembler of the AMAZE system does a syntactic analysis of Table 3.6 and generates the fuse map file. Finally the Device Program Interface of AMAZE sends this file to the device programmer and the PLS100 is programmed.

3.4.2 Sequential logic controller design using PLDs

The state diagram is a good method of specifying a sequential system in order to implement the system using PLDs. As was seen in section 1.3.2 of chapter 1, the state diagram can be defined using levels or edges of the input variables. In most practical situations, the state diagram can be obtained directly from the specifications of input variables. Figure 3.9 illustrates specifications using the logic product of x_i input variables during two successive edges of the clock. Several examples are now studied.

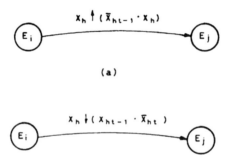

FIGURE 3.9 *Input variable edges expressed by means of logic products.*

EXAMPLE 3.2

Implementation of the synchronous logic controller shown in example 1.2, using PLDs.

Solution

The state diagram of Figure 1.17 is obtained from the specifications of example 1.2. It is also illustrated in Figure 3.10 indicating the direct or inverted variable levels that cause the transition between states.

The 5C060 device from Intel (described in appendix 3) is used to implement this controller. Table 3.7 shows the controller specification in a high level language compatible with the IPLSII program system. In the INPUT part, the two input variables A and B and the clock (CLK) are included. In the OUTPUT part, the internal variable states Q0, Q1 and Q2, and the external output Z are included.

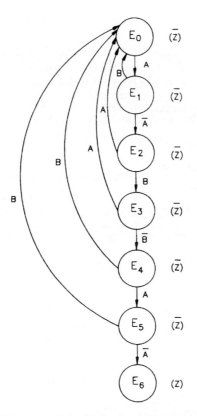

FIGURE 3.10 *State diagram of the logic controller for the lock of example 3.2 specified by input variable levels.*

In the NETWORK part, the macrocells that generate the synchronized inputs A_t and B_t from A and B are defined. These macrocells must be programmed as type D flip-flop without external output [no output, registered feedback (NORF)] (Figure 3.11a). There is also another macrocell assigned to the output Z; this macrocell must be programmed as combinational output without feedback [combinational output, no feedback (CONF)] (Figure 3.11b).

In the STATES part, the state code assignment is done in binary. The transitions between states are expressed using high-level language statements.

By running the IPLSII program we obtain the minimized logic equations of the internal state variables and the output variables. The results are given in Table 3.8; that in the NETWORK part includes the specification of all variables as follows:

1. Three normal inputs INP for inputs A, B and the clock.
2. One RORF (registered output, registered feedback) macrocell for Q0 internal state variable.

TABLE 3.7

Department of Electronic Technology
University of Vigo (Spain)
Synchronous Logic Controller of an Electronic Lock
(activated by input variable levels)

PART : 5C060

INPUTS : A,B,CLK

OUTPUTS : Q2,Q1,Q0,Z

NETWORK:
At = NORF(A,CLK,GND,GND)
Bt = NORF(B,CLK,GND,GND)
Z = CONF(Z,VCC)

MACHINE : LOCK
          CLOCK : CLK

```
STATES:      [ Q2   Q1   Q0]
E0           [ 0    0    0 ]
E1           [ 0    0    1 ]
E2           [ 0    1    0 ]
E3           [ 0    1    1 ]
E4           [ 1    0    0 ]
E5           [ 1    0    1 ]
E6           [ 1    1    0 ]
```

E0: IF At THEN E1

E1: IF At' THEN E2
    IF Bt THEN E0

E2: IF Bt THEN E3
    IF At THEN E0

E3: IF Bt' THEN E4
    IF At THEN E0

E4: IF At THEN E5
    IF Bt THEN E0

E5: IF At' THEN E6
    IF Bt THEN E0

E6: IF At*At' THEN E6
    ASSERT:
              Z

END$

LOGIC CONTROLLER DESIGN USING PLDS

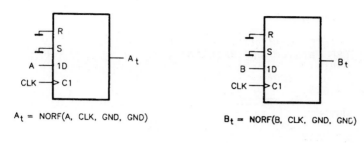

A_t = NORF(A, CLK, GND, GND)

B_t = NORF(B, CLK, GND, GND)

(a)

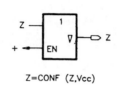

Z=CONF (Z,Vcc)

(b)

FIGURE 3.11 *IPLSII symbols of macrocells that generate input synchronized variables A_t and B_t* (a) *and output variables* (b) *in example 3.2.*

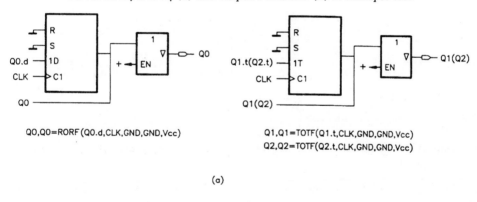

Q0,Q0=RORF(Q0.d,CLK,GND,GND,Vcc)

Q1,Q1=TOTF(Q1.t,CLK,GND,GND,Vcc)
Q2,Q2=TOTF(Q2.t,CLK,GND,GND,Vcc)

(a)

CLK=INP(CLK)

A=INP(A)

B=INP(B)

(b)

FIGURE 3.12 *IPLSII symbols of macrocells used to implement internal state variables* (a) *and symbols of input variables CLK, A and B* (b) *in example 3.2.*

TABLE 3.8

```
iSTATE Version 1.5 Revision 3.0
LEF Version 1.5 Baseline 3.3i
PART:
        5C060
INPUTS:
        A, B, CLK
OUTPUTS:
        Q2, Q1, Q0, Z, At, Bt
NETWORK:
        CLK = INP(CLK)
        A = INP(A)
        B = INP(B)
        Z = CONF(Z, VCC)
        Q0, Q0 = RORF(Q0.d, CLK, GND, GND, VCC)
        Q1, Q1 = TOTF(Q1.t, CLK, GND, GND, VCC)
        Q2, Q2 = TOTF(Q2.t, CLK, GND, GND, VCC)
        At = NORF(A, CLK, GND, GND)
        Bt = NORF(B, CLK, GND, GND)
EQUATIONS:
        Q2.t = Q2 * Q1' * Q0' * Bt * At'
             + Q2' * Q1 * Q0 * Bt'
             + Q2 * Q1' * Q0 * Bt * At;

        Q1.t = Q2' * Q1 * Q0 * Bt'
             + Q2' * Q1 * At * Bt'
             + Q2' * Q1 * Q0 * At
             + Q1' * Q0 * At';

        Q0.d = Q2' * Q1 * Q0' * Bt
             + Q2' * Q1 * Bt * At'
             + Q1' * Bt'* At
             + Q1' * Q0 * At;

        Z = Q2 * Q1 * Q0';

END$
```

3. Two TOTF (T output, T feedback) macrocells, one for every Q1 and Q2 internal state variables.

The logic symbols of the different macrocells are shown in Figure 3.12. The IPLSII specifies the value of set and reset inputs of the macrocell, although the PLD used (the 5C060) does not have a reset input.

Next, the IPLSII system assigns the input and output variables to the 5C060 pins (Figure 3.13). Finally the IPLSII generates a file that is sent to the device programmer.

The design can also be implemented with the PLS157 device from Signetics using the AMAZE CAD system.

We first assign the device pins to the input, output and internal state variables (the last for observation purposes only). The results are shown in Figure 3.14. The PLS157 device has four input pins (I) and twelve bidirectional pins (/B) than can be

LOGIC CONTROLLER DESIGN USING PLDS

```
INTEL Logic Optimizing Compiler Utilization Report        LOCK.rpt
FIT Version 1.5 Level 3.0i

***** Design implemented successfully


Department of Electronic Technology
University of Vigo (Spain)
Synchronous Logic Controller of an Electronic Lock
(activated by input variable levels)

iSTATE Version 1.5 Revision 3.0

                   5C060
                   - - - -
         CLK -|  1     24|- Vcc
           B -|  2     23|- Gnd
    RESERVED -|  3     22|- Gnd
    RESERVED -|  4     21|- Gnd
          Q2 -|  5     20|- Gnd
          Q1 -|  6     19|- Gnd
          Q0 -|  7     18|- Gnd
           Z -|  8     17|- Gnd
         Gnd -|  9     16|- Gnd
         Gnd -| 10     15|- Gnd
           A -| 11     14|- Gnd
         GND -| 12     13|- Gnd
                   - - - -
```

FIGURE 3.13 *PLD 5C060 pin assignment to the input and output variables of logic controller for the lock of example 3.2.*

```
File Name : lock
Date :
Time :

#################### P I N   L I S T ####################

     LABEL    ** FNC **PIN - - - - - PIN** FNC **    LABEL
      N/C     ** CK  **  1-|         -20 ** +5V  **VCC
        A     ** I   **  2-|         -19 ** /B   **N/C
        B     ** I   **  3-|  P      -18 ** /B   **N/C
    RESET     ** I   **  4-|  L      -17 ** /B   **N/C
      N/C     ** I   **  5-|  S      -16 ** /O   **Z
      N/C     ** /B  **  6-|  1      -15 ** /O   **E2
      N/C     ** /B  **  7-|  5      -14 ** /O   **E1
      N/C     ** /B  **  8-|  7      -13 ** /O   **E0
      N/C     ** /B  **  9-|         -12 ** /B   **N/C
      GND     ** 0V  ** 10-|         -11 ** /OE  **N/C
                              - - - -
```

FIGURE 3.14 *PLD PLS157 pin assignment to the input and output variables of the logic controller for the lock of example 3.2.*

TABLE 3.9

```
@DEVICE SELECTION
LOCK / PLS157

@STATE VECTORS
[E2,E1,E0]
ESTIN = ---b;      " - : don't care "
EST0  = 111b;
EST1  = 110b;
EST2  = 101b;
EST3  = 100b;
EST4  = 011b;
EST5  = 010b;
EST6  = 001b;

@INPUT VECTORS
[A,B]
INDET = --b;
IN0 = 00b;
IN1 = 01b;
IN2 = 10b;
A1  = 1-b;
B1  = -1b;

@OUTPUT VECTORS
[Z]
OUT0' = 1b;
OUT1' = 0b;

@TRANSITIONS
While [est0]
        If [in2] then [est1] with [out0']
While [est1]
        case
        [in0]::[est2] with [out0']
        [B1]::[est0] with [out0']
        endcase
While [est2]
        case
        [in1]::[est3] with [out0']
        [A1]::[est0] with [out0']
        endcase
While [est3]
        case
        [in0]::[est4] with [out0']
        [A1]::[est0] with [out0']
        endcase
While [est4]
        case
        [in2]::[est5] with [out0']
        [B1]::[est0] with [out0']
        endcase
While [est5]
        case
        [in0]::[est6] with [out1']
        [B1]::[est0] with [out0']
        endcase
While [est6]
        if [indet] then [est6] with [out1']
```

LOGIC CONTROLLER DESIGN USING PLDS   111

programmed as inputs or outputs (see appendix 3). In this example, four are programmed as outputs (three for internal state variables and one for the Z output that controls the lock).

Since state E_6 in Figure 3.10 does not have an exit, an asynchronous reset input (pin 4) is used to return the controller to the initial state by touching a push-button.

Next, we use the Edit State Transfer Option of the AMAZE system to define the state vectors and the specification in a structured high-level language. Table 3.9 shows the result.

Finally, the AMAZE assembler program does the syntactic analysis and generates the fuse map file that is sent to the device programmer.

EXAMPLE 3.3

Implementation of the logic controller described in example 1.5 using PLDs.

Solution

Figure 3.15 shows the state diagram (same as in Figure 1.26). Expressing the input edges by logic products we describe the specifications in the MACHINE part of Table 3.10.

In the NETWORK part of Table 3.10 we specify a CONF (combinational output, no feedback) macrocell for the Z output, and four NORF (no output, registered

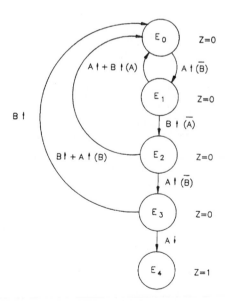

FIGURE 3.15 *State diagram of the logic controller of example 3.2 specified by input variable edges.*

TABLE 3.10

Department of Electronic Technology
University of Vigo (Spain)
Synchronous Logic Controller of an Electronic Lock
(activated by input variable edges)

PART : 5C060

INPUTS : A,B,CLK

OUTPUTS : Q2,Q1,Q0,Z

NETWORK:
At=NORF(A,CLK,GND,GND)
Bt=NORF(B,CLK,GND,GND)
At_1=NORF(At,CLK,GND,GND)
Bt_1=NORF(Bt,CLK,GND,GND)
Z=$\overline{\text{CONF}}$(Z,VCC)

MACHINE : LOCK_EDGES
          CLOCK : CLK

STATES: [Q2   Q1   Q0]
E0   [0   0   0]
E1   [0   0   1]
E2   [0   1   0]
E3   [0   1   1]
E4   [1   0   0]

E0: IF (Bt'*(At*At_1')) THEN E1

E1: IF ((At*(Bt*Bt_1'))+(At*At_1'))   THEN E0
    IF (At'*(Bt*Bt_1')) THEN $\overline{\text{E2}}$

E2: IF ((Bt*(At*At_1'))+(Bt*Bt_1'))   THEN E0
    IF (Bt'*(At*At_1')) THEN $\overline{\text{E3}}$

E3: IF (Bt*Bt_1')   THEN E0
    IF (At'*At_1)   THEN E4

E4: IF At_1*At_1'   THEN E4
    ASSER$\overline{\text{T}}$:
            Z

END$

TABLE 3.11

```
Department of Electronic Technology
University of Vigo (Spain)
Synchronous Logic Controller of an Electronic Lock
(activated by input variable edges)

iSTATE Version 1.5 Revision 3.0
LEF Version 1.5 Baseline 3.3i
PART:
    5C060
INPUTS:
    A, B, CLK
OUTPUTS:
    Q2, Q1, Q0, Z, At, Bt, At_1, Bt_1
NETWORK:
    CLK = INP(CLK)
    A = INP(A)
    B = INP(B)
    Z = CONF(Z, VCC)
    Q0, Q0 = RORF(Q0.d, CLK, GND, GND, VCC)
    Q1, Q1 = TOTF(Q1.t, CLK, GND, GND, VCC)
    Q2, Q2 = TOTF(Q2.t, CLK, GND, GND, VCC)
    At = NORF(A, CLK, GND, GND)
    Bt = NORF(B, CLK, GND, GND)
    At_1 = NORF(At, CLK, GND, GND)
    Bt_1 = NORF(Bt, CLK, GND, GND)
EQUATIONS:
    Q2.t = Q2' * Q1 * Q0 * At' * At_1 * Bt_1
         + Q2' * Q1 * Q0 * At' * At_1 * Bt';

    Q1.t = Q2' * Q1 * Q0' * Bt * At * At_1'
         + Q2' * Q0 * Bt * Bt_1' * At'
         + Q2' * Q1 * Q0 * At' * At_1
         + Q2' * Q1 * Bt * Bt_1';

    Q0.d = Q2' * Q1 * Q0 * At_1' * Bt'
         + Q2' * Q0 * At * At_1 * Bt'
         + Q2' * Q1' * Q0 * At'* Bt'
         + Q2' * Q1 * Q0 * At_1' * Bt_1;
         + Q2' * Q0 * At * At_1 * Bt_1
         + Q2' * Q1' * Q0 * At'* Bt_1
         + Q2' * Q0' * At * At_1' * Bt';

    Z = Q2 * Q1' * Q0';

END$
```

```
INTEL Logic Optimizing Compiler Utilization Report      LOCK_EDGES.rpt
FIT Version 1.5 Level 3.0i

***** Design implemented successfully


Department of Electronic Technology
University of Vigo (Spain)
Synchronous Logic Controller of an Electronic Lock
(activated by input variable edges)

iSTATE Version 1.5 Revision 3.0

                    5C060
                    - - - - -
         CLK  -|  1     24 |- Vcc
           B  -|  2     23 |- Gnd
    RESERVED  -|  3     22 |- Gnd
    RESERVED  -|  4     21 |- Gnd
    RESERVED  -|  5     20 |- Gnd
    RESERVED  -|  6     19 |- Gnd
          Q0  -|  7     18 |- Gnd
          Q1  -|  8     17 |- Gnd
          Q2  -|  9     16 |- Gnd
           Z  -| 10     15 |- Gnd
           A  -| 11     14 |- Gnd
         GND  -| 12     13 |- Gnd
                    - - - - -
```

FIGURE 3.16 *PLD 5C060 pin assignment to the input and output variables of the logic controller for the lock of example 3.3.*

feedback) macrocells for the inputs. Two of them generate the synchronized input variables A_t and B_t, and the other two generate the delayed input variables A_{t-1} and B_{t-1}.

In the MACHINE part of Table 3.10 the internal states are defined and the transitions between them are expressed in a high-level language.

The IPLSII system uses the information of Table 3.10 to generate the minimized equations of internal state and output variables. The results are shown in Table 3.11.

Figure 3.16 indicates the assignment of the 5C060 pins to the input and output variables as generated by IPLSII.

EXAMPLE 3.4

Implementation of the logic controller described in example 1.4 using PLDs.

Solution

Figure 3.17 shows the state diagram (the same as in Figure 1.27). Since it is a very simple logic controller, it is possible to obtain without CAD the minimum PLD device structure necessary to implement it. The internal states E_1 and E_2 can be implemented with a J–K flip-flop that is at 0 when the controller is in the E_1 state and at 1 when it is in the E_2 state.

LOGIC CONTROLLER DESIGN USING PLDS

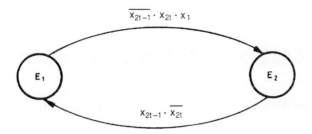

FIGURE 3.17 *State diagram of the logic controller for the bar selection system, with the input variable edges expressed by means of logic products.*

To achieve the change of the flip-flop from 0 to 1, the J input must be at 1 and the K input at 0. In the same way, the J input must be at 0 and the K at 1 to change the flip-flop from 1 to 0. From these conditions and the state diagram of Figure 3.17 we obtain the following logic equations:

$$J = \overline{x_{2t-1}} x_{2t} x_1 \overline{Q}$$
$$K = x_{2t-1} \overline{x_{2t}} Q$$

TABLE 3.12

```
Department of Electronic Technology
University of Vigo (Spain)
Synchronous Logic Controller of a Bar Selection System

PART : 5C060

INPUTS : X1, X2, CLK

OUTPUTS : Z
NETWORK:
X1t=NORF(X1, CLK, GND, GND)
X2t=NORF(X2, CLK, GND, GND)
X2t_1=NORF(X2t, CLK, GND, GND)

MACHINE : BAR_SELECTION
          CLOCK : CLK

STATES :      [ Z ]
E1            [ 0 ]
E2            [ 1 ]

E1 :
    IF (X2t_1'*X2t*X1t) THEN E2

E2 :
    IF (X2t_1*X2t') THEN E1

END$
```

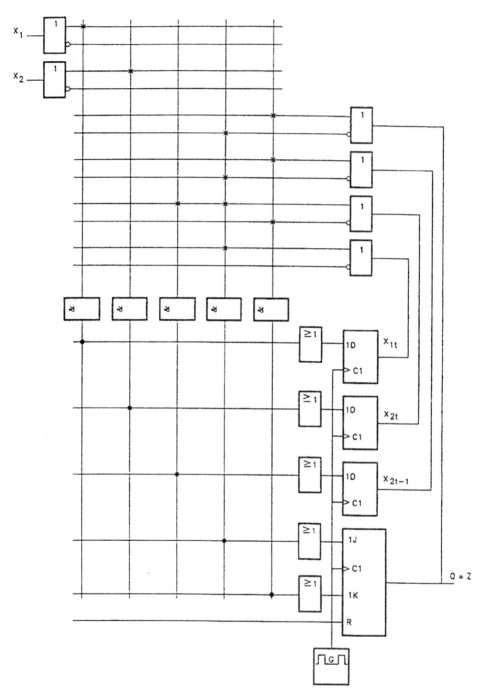

FIGURE 3.18 *Minimum configuration of a PLD that implements the logic controller for the bar selection system of example 3.4.*

TABLE 3.13

```
Department of Electronic Technology
University of Vigo (Spain)
Synchronous Logic Controller of a Bar Selection System

iSTATE Version 1.5 Revision 3.0
LEF Version 1.5 Baseline 3.3i

PART :
        5C060

INPUTS:
        X1, X2, CLK

OUTPUTS:
        Z, X1t, X2t, X2t_1

NETWORK:
        CLK = INP(CLK)
        X1 = INP(X1)
        X2 = INP(X2)
        Z, Z = TOTF(Z.t, CLK, GND, GND, VCC)
        X1t=NORF(X1, CLK, GND, GND)
        X2t=NORF(X2, CLK, GND, GND)
        X2t_1=NORF(X2t, CLK, GND, GND)

EQUATIONS:
        Z.t = Z' * X2t_1' * X2t * X1t
            + Z * X2t_1 * X2t';

END$


  INTEL Logic Optimizing Compiler Utilization Report
  FIT Version 1.5 Level 3.0i

  ***** Design implemented successfully


  Department of Electronic Technology
  University of Vigo (Spain)
  Synchronous Logic Controller of Bar Selection System

  iSTATE Version 1.5 Revision 3.0

                    5C060
                  - - - - -
          CLK  - | 1     24 | - Vcc
           X2  - | 2     23 | - Gnd
     RESERVED  - | 3     22 | - Gnd
     RESERVED  - | 4     21 | - Gnd
     RESERVED  - | 5     20 | - Gnd
            Z  - | 6     19 | - Gnd
          Gnd  - | 7     18 | - Gnd
          Gnd  - | 8     17 | - Gnd
          Gnd  - | 9     16 | - Gnd
          Gnd  - |10     15 | - Gnd
           X1  - |11     14 | - Gnd
          GND  - |12     13 | - Gnd
                  - - - - -
```

FIGURE 3.19 *PLD 5C060 pin assignment to the input and output variables of the logic controller for the bar selection system of example 3.4.*

Figure 3.18 shows the minimum PAL-based PLD hardware. Two D flip-flops synchronize the X_1 and X_2 variables and one more generates X_{2t-1}.

We can implement this controller with the 5C060 PLD. We first describe the controller according to the IPLSII rules (Table 3.12). The IPLSII generates the minimized equations and assigns one macrocell to every variable x_{1t}, x_{2t}, CLK and Z. The results are shown in Table 3.13. Finally the IPLSII does the 5C060 pin assignment. The result is shown in Figure 3.19.

Logic controllers of examples 3.2, 3.3 and 3.4 are implemented using the 5C060 PLD from Intel. Some macrocells of the 5C060 are used to synchronize the input variables and to generate the delayed ones (at time $t-1$). This PLD, though, is misused, especially when there are a large number of input variables defined, because there are a large number of AND and OR gates not being used. For example, in the logic controller of example 3.4 (Table

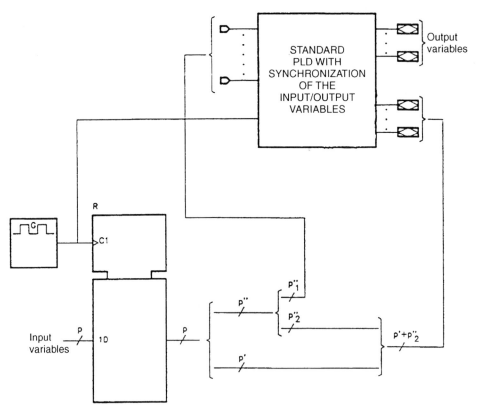

FIGURE 3.20 *Logic controller implemented with a standard PLD and a register that synchronizes the input variables.*

3.11), eight macrocells are used although the controller has only three internal state variables and one output variable. Several more efficient solutions to logic controller implementation using PLDs are possible:

1. If the PLD has input/output pins that can be fed back to the AND gate array through flip-flops (when they are programmed as inputs), the circuit of Figure 3.20 can be used. This circuit has an external register R to synchronize all p input variables. The outputs of this register are connected to the PLD as follows:
   - the p'' level defined input variables are connected to input pins (p''_1) or to bidirectional pins (p''_2);
   - the p' edge defined input variables are connected to bidirectional pins.

2. When the bidirectional PLD pins cannot be connected to the AND gate array through flip-flops, we use external registers R_1 and R_2 to synchronize the input variables and to generate the delayed ones (input variables at time $t-1$) (Figure 3.21).

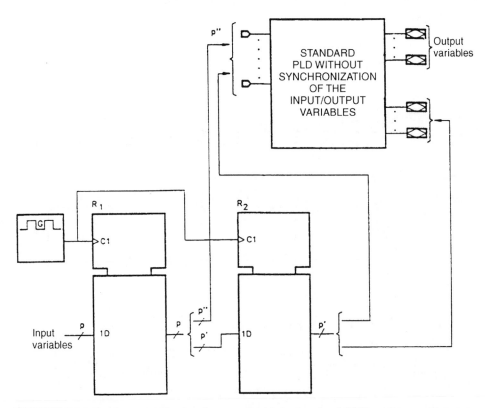

FIGURE 3.21 *Logic controller implemented with a standard PLD and two registers to synchronize and delay the input variables.*

3. Using a PLD orientated to the implementation of logic controllers [PERE 91] whose internal state changes are specified by input variable edges. This PLD must have the following characteristics:
    - hardware with all the elements necessary to synchronize the input variables and generate the delayed ones;
    - input variables that can be defined by levels or by edges;
    - input/output bidirectional pins to allow for the implementation of logic controllers with a different number of input and output variables.

Figure 3.22 shows a PLD with these characteristics. This circuit has a buried register R_1 with n_1 edge-triggered D flip-flops. The Q outputs of this register are connected to the input array. The number n_1 of flip-flops can be twice the number n of input variables in order to use the complete circuit capacity when all the variables are specified by means of edges.

The buried register outputs are also connected to an AND gate array. The AND gates are divided into p groups, and each one is connected to an OR gate to constitute a PAL. The output of every OR gate is connected to an edge-triggered D flip-flop (rise edge in Figure 3.22).

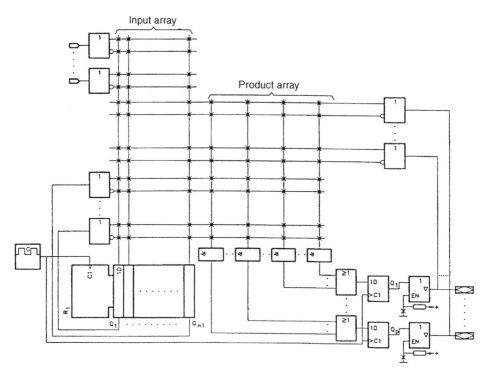

FIGURE 3.22 *Hardware architecture of a PLD orientated to the implementation of logic controllers specified by input variable edges.*

In order to be able to use this PLD for logic controller implementation when the number of input variables is greater than the number of input pins, the D flip-flops are connected to tri-state gates, and the logic value of the enable inputs (EN) can be programmed.

Bibliography

[INTE 88]   Intel Corporation, *IPLSII User's Guide*, 1988.
[PERE 91]   S.A. Pérez, 'New systematic methods for logic controller design', PhD thesis, Department of Electronic Technology, University of Vigo, 1991.
[SIGN 87]   Signetics Corporation, *Software for Programmable Logic Devices*, 1987.

PART 3
Programmable logic controllers

In this part we study programmable and modular logic controllers which have a logic unit, and are usually called PLCs (programmable logic controllers).

These systems are modular and easy to program. Therefore they are suited to the control of medium- and large-sized industrial processes which normally change the number of input and output variables during the lifetime of the logic controller.

PLCs are standard off-the-shelf products and their programs are easy to copy, so they are suitable for applications where the copyright resides on the controlled process.

CHAPTER 4

Introduction to programmable logic controllers

4.1 Introduction

In chapter 1 we looked at logic controllers that execute logic operations by jumping between internal states of a register or counter. These logic controllers execute the algorithm corresponding to a combinational or sequential system, without using a logic operating unit. A feature of such systems is that their instructions (Figures 1.42 and 1.44) give no clear indication about the operation they execute because they have no operation field.

In this chapter, programmable logic controllers with a logic unit are studied. These systems execute logic equations by means of a sequence of logic operations carried out by a combinational circuit that constitutes a logic unit. For the same reasons as indicated in section 1.3.2, these systems are of interest only if they are implemented using a programmable combinational circuit that contains the instructions.

Modular programmable logic controllers with a logic unit are used, mainly in their minimal versions, in the implementation of sequential control systems activated by the state or the change of state of external input variables. In this role they act just like a logic controller and consequently they are called programmable logic controllers (PLCs).

Figure 4.1 shows the diagram of a basic PLC made up of two parts:

1. A control unit with a clock, a synchronous counter, a passive memory [ROM, PROM or RPROM (EPROM, EEPROM or FLASH)] and a combinational circuit which generates the control signals of the operating unit.
2. An operating unit with an input unit (IU), a logic unit (LU), a flip-flop to store partial results [result flip-flop (RF)] and an output unit (OU).

The logic unit must be able to execute the logic AND and OR functions between input variables or their complements. The operating unit control

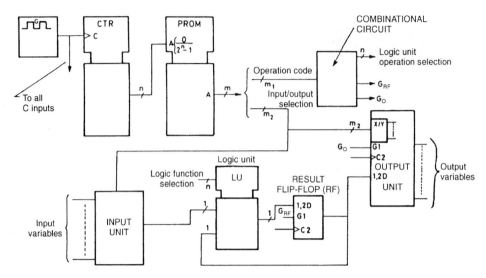

FIGURE 4.1 *Block diagram of a basic PLC.*

signals are obtained from the operation code and they are the following:

1. A combination of m_1 bits which select a maximum of 2^{m1} different operations that can be executed by the operating unit.
2. A variable G_{RF} which enables the storage of the result in RF when its logic state is 1.
3. An enable variable of the output unit G_O which enables the transfer of the content of the result flip-flop (RF) to the output unit flip-flop selected by means of the input/output selection field of the PROM memory.

In Figure 4.1 and thereafter, the new standard symbols are used. The reader not familiar with these symbols is advised to consult appendix 1.

This system differs from the one studied in section 1.3.2 in the format of the binary output combination of the PROM memory. The reader can compare Figure 4.2, representing the output vector of the PROM memory of

OPERATION CODE	INPUT/OUTPUT SELECTION CODE

OPERATION CODE { Specifies the operation to be performed by the logic unit

INPUT/OUTPUT SELECTION CODE { Selects the input or the output variable

FIGURE 4.2 *Output vector of the PROM memory of the PLC of Figure 4.1.*

INTRODUCTION TO PLCS

Figure 4.1, with Figure 1.42 representing the vector of the combinational circuit of Figure 1.39.

The vector of Figure 4.2 is divided into two parts known as **fields**:

1. The operation code field which specifies the operation to be performed by the logic unit.
2. The input/output selection field, which selects either the input variable to be used in the operation indicated by the operation code field or the output flip-flop to which information is transferred.

The system of Figure 4.1 must fulfil the requirement that by executing a sequence of instructions, starting from the zero location of the PROM memory, any logic function specified by a sum-of-products or a product-of-sums can be performed.

Analyzing the system hardware of Figure 4.1, we see that logic products or logic sums can be performed but not combinations of both. In fact, let us assume that there is a combination of the operation code that executes the following actions:

1. Selects the logic AND operation by means of the logic unit operation selection variables.
2. Sets to value 1 the enable input G_{RF} of the result flip-flop.

When the instruction located in the PROM memory position addressed by the counter has in the operation field the binary combination which executes the actions indicated above, the system stores the logic product of the content of RF and the selected input variable in the result flip-flop (RF).

Therefore, if the RF flip-flop is initially at state 1, the system stores in RF the value of the logic product of certain input variables by executing a sequence of AND instructions. However, to execute a logic sum-of-products it is necessary to store one of them while the other is being calculated in order to execute their logic sum, and this is not possible using the system of Figure 4.1.

On the other hand, the implementation of sequential systems implies that the PLC must be able to decide whether or not to execute certain actions, but this is not possible with the system of Figure 4.1.

If the circuit diagram of Figure 4.1 is changed slightly, we can obtain a PLC capable of executing a sequence of instructions which simulate any combinational or sequential system. There are several ways to achieve these changes. That is why different PLCs with different hardware in their operating unit can be implemented, which in turn gives rise to a different set of instructions [AN 76] [MOTO 77].

In the following sections, two practical solutions are analyzed. These solutions show that it is possible to conceive a basic PLC in many different ways.

4.2 Basic PLC with load and store instructions

Figure 4.3 shows a PLC which allows the execution of a sum-of-products by temporarily storing the products in auxiliary flip-flops. In this system, the output unit flip-flops are used for that purpose. The output unit is also a temporary memory unit (OU/T) whose contents can be transmitted to one of the logic unit inputs. This is achieved by connecting the output unit flip-flops to the input of the logic unit via tri-state gates. Every output unit flip-flop can be used to store either output variables or internal state variables.

The operating unit of this PLC has not only signals G_{RF} and G_O of the basic PLC of Figure 4.1, but also the following control signals:

1. The enable signal EN_I of the tri-state output of the input unit. This signal allows the selected input variable to be transmitted to bus B connected to the input of the logic unit.
2. The enable signal EN_{RF} of the tri-state output of the result flip-flop (RF). This signal allows the content of RF to be transmitted to bus B.
3. The enable signal EN_T of the tri-state output of the output/temporary unit (OU/T). This signal allows the content of the selected flip-flop to be transmitted to bus B.

It is assumed that the system has the instructions indicated in Table 4.1. This table also specifies the actions corresponding to each instruction. On the right of the table we can see the actions in the following symbolic language:

Action	Symbolic representation
An input variable in the bus D	$IU \rightarrow D$
Transfer of bus D content to flip-flop RF	$(D) \rightarrow RF$
Transfer of the inverted content of bus D to flip-flop RF	$(\bar{D}) \rightarrow RF$
Logic AND of the content of RF and bus D and storage of the result in RF	$(RF) \cdot (D) \rightarrow RF$

where D = bit bus connected to the output of the input unit, RF = result flip-flop, (RF) = content of the result flip-flop, IU = input unit and OU/T = output and temporary storage unit.

The output and temporary storage unit (OU/T) of Figure 4.3 deserves special attention. This unit consists of 2^{m2} flip-flops with the D (1, 2D) input connected to bus B and input C (C2) connected to the pulse generator (these connections are indicated by symbols Z3 and Z4 respectively). Each flip-flop has also a variable G1, which is the logic product of an output of the selection decoder and signal G_O. Furthermore, each flip-flop has an output to the outside world which constitutes an output variable and another internal tristate output connected to bus B. This output is enabled by signal EN_T.

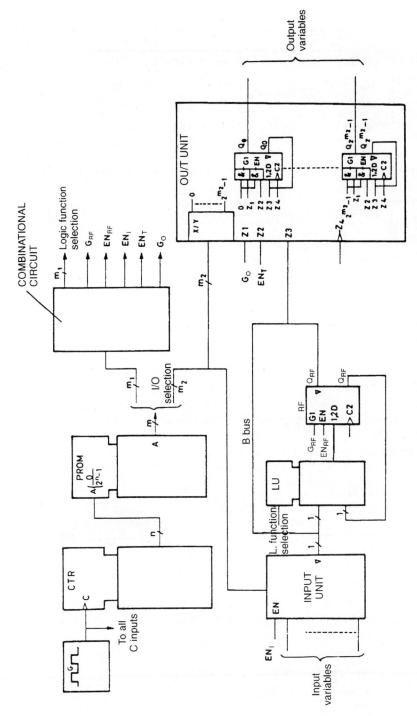

FIGURE 4.3 Block diagram of a basic PLC with load and store instructions.

TABLE 4.1

Instruction	Actions
LOAD	IU → D and (D) → RF
INVERTED LOAD	IU → D and ($\overline{D}$) → RF
AND	IU → D and (RF)(D) → RF
INVERTED AND	IU → D and (RF)($\overline{D}$) → RF
OR	OU/T → D and (RF)+(D) → RF
INVERTED OR	OU/T → D and (RF)+($\overline{D}$) → RF
STORE	(RF) → OU/T
INVERTED STORE	($\overline{RF}$) → OU/T

D   Bit bus connected to the output of the input unit
RF   Result flip-flop
(RF) Content of result flip-flop
IU   Input unit
OU/T Output and temporary storage unit

This PLC can execute any logic sum-of-products. To prove it, we assume that the binary combinations assigned to each instruction of Table 4.1 are those indicated in Table 4.2.

Let $f = ab + \bar{c}d$ be the function to be executed. We assign to the input variables the following octal selection combinations: $a = 100_8$, $b = 101_8$, $c = 102_8$, $d = 103_8$. We also assign to function f the flip-flop of the output unit selected by the combination 50_8. It is also assumed that combination 51_8 selects a flip-flop used as a temporary memory. Table 4.3 shows the instruction sequence which executes f.

TABLE 4.2

Instruction	Code	
	Binary	Octal
LOAD	000	0
INVERTED LOAD	001	1
AND	010	2
INVERTED AND	011	3
OR	100	4
INVERTED OR	101	5
STORE	110	6
INVERTED STORE	111	7

TABLE 4.3

PROM memory address (Octal)	Instruction	Comment
000	0.100	Transfers variable a to RF
001	2.101	AND of a and b to RF
002	6.051	Stores a b
003	1.102	Transfers $\bar{c}$ to RF
004	2.103	AND of $\bar{c}$ and d to RF
005	4.051	OR of a b and $\bar{c}$ d → RF
006	6.050	Transfers (RF) to the output

However, the system of Figure 4.3 is not able to decide whether to execute certain actions. Several solutions exist to overcome this problem, among which are the following:

1. To endow the system with jump instructions which have a jump address in the second field instead of a selection combination of an input or output variable. The content of the counter is incremented by 1 if flip-flop RF is at 0; on the other hand, if flip-flop RF is at 1, the jump address is loaded, in parallel, into the counter. Therefore, we have a PLC with the two types of instruction indicated in Figure 4.4.

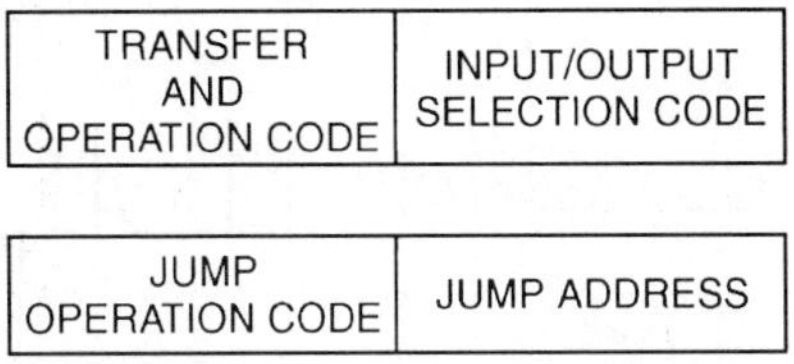

TRANSFER AND OPERATION CODE	INPUT/OUTPUT SELECTION CODE

JUMP OPERATION CODE	JUMP ADDRESS

FIGURE 4.4 *Format of the instructions of a PLC with jump instructions.*

To make this possible, it is necessary to change the hardware of Figure 4.3 as indicated in Figure 4.5. In this figure the counter is endowed with an operation mode input (M1) and parallel inputs connected to the address field of the PROM memory. Depending on whether M1 is 0 or 1, the counter increments its contents or receives the information from the parallel inputs, respectively. The input M1 is equal to the logic product:

$M1 = Q_{RF} \cdot I_J$

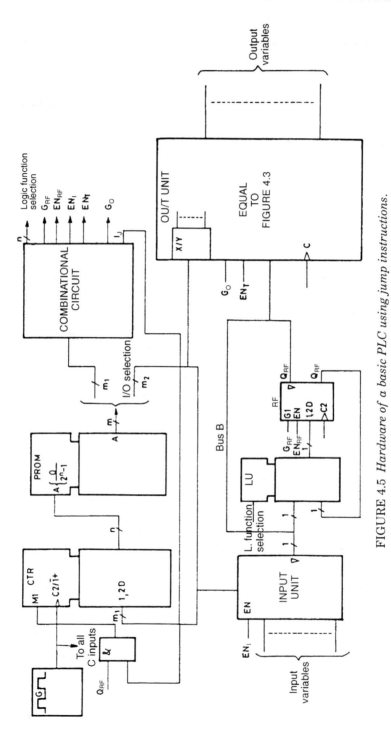

FIGURE 4.5 *Hardware of a basic PLC using jump instructions.*

INTRODUCTION TO PLCS   133

where Q_{RF} is the output of the flip-flop RF, and I_J is the output of the combinational circuit. This signal is a logic 1 if the operation code corresponds to the jump instruction. In this way, if $I_J = 0$ the counter increments its contents, and if $I_J = 1$ the counter increments or jumps according to Q_{RF} being 0 or 1, respectively.

2. To endow the system with two instructions, one of conditional disabling (CONDIS) of instruction execution and another of unconditional enabling (UNCENA) of such execution. Disabling is achieved by ensuring that the control unit remains in a state where all the control signals of all the elements of the operating unit are switched off. To achieve this, we add to the control unit a disabling flip-flop (Q_{DIS}) which is set to one if, when selecting a CONDIS instruction, the content of RF is 0, and is set to 0 otherwise.

    Enabling is achieved by ensuring that Q_{DIS} returns to 0 when an UNCENA instruction is executed.

    The reader is invited to determine the modifications necessary in Figure 4.3 to achieve this operation mode.

3. The instructions which transfer information to the output flip-flops may or may not be executed, depending on the result of the last operation performed. In turn, this way of decision-making has two variants:
    - to endow the system with an output disabling flip-flop ODF and an instruction that transfers the content of RF to the output flip-flop;
    - to endow the system with conditional operating instructions, such as set or reset the selected output flip-flop as a function of the state of RF.

Figure 4.6 shows the alternative to endow the PLC of Figure 4.3 with a disabling flip-flop. An inhibition input is added to the output unit. This input is connected to the output of a flip-flop named 'output disabling flip-flop' (ODF) whose input is connected to the output of the result flip-flop. The system must be endowed with an instruction that transfers the content of RF to ODF.

If an instruction to transfer the content of RF to ODF is stored in the PROM memory and is followed by an output instruction, it will or will not act upon the content of the corresponding flip-flop according to whether the ODF is a logic 0 or 1, respectively.

Any of the proposed solutions is feasible and gives rise to a different PLC.

4.3 Basic PLC with conditional operating instructions

The operating unit of the PLC described in the previous section has a single flip-flop for the temporary storage of partial results.

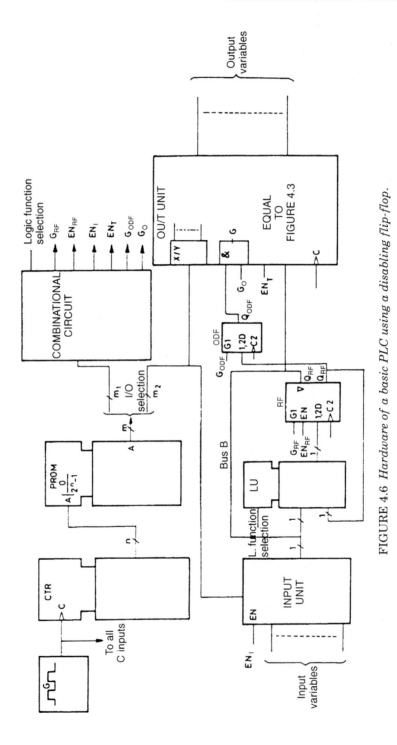

FIGURE 4.6 *Hardware of a basic PLC using a disabling flip-flop.*

INTRODUCTION TO PLCS

The use of two flip-flops RF1 and RF2 in the operating unit (Figure 4.7) allows the execution of a sum-of-products by a sequence of only logic OR and AND instructions. To show this, we assume the following:

1. Initially the contents of RF1 and RF2 are, respectively, 1 and 0.
2. The logic AND instruction executes the logic AND operation of the input variable, specified by the selection field and the content of RF1, and stores the result in RF1.
3. The logic OR instruction executes two actions simultaneously:
   - the logic OR of the contents of RF2 and RF1 and the storage of the result in RF2;
   - loading RF1 with the input variable selected in the input selection field.

The PLC has an equality instruction executing the logic OR of the content of RF1 and RF2 and the storage of the result in the selected output variable.

Figure 4.7 only shows the hardware of the operating part which allows the execution of all of the described actions. There we can see buses B1 and B2. B1 is formed by an output of the input unit and an output of the logic unit (LU). The second, B2, is formed by the output of RF2 and the other output of the input unit. We invite the reader to deduce which signals are to be activated at each time in order to execute the operations indicated above.

To analyze how a sum of products is executed we assume the PLC has the instructions indicated in Table 4.4. Note that this PLC does not have load and store instructions because they are not necessary.

The following example shows the way of executing the expression $ab + cd$ using this PLC. If the input variables a, b, c and d are assigned to the selection combinations 100, 101, 102 and 103 in octal, the instructions to be sequentially executed are:

    0.100
    0.101
    3.102
    0.103

Just like in the PLC of the previous section, in order to implement a sequential system there must be conditional instructions that provide decision-making capabilities as a function of the state or the change of state of some input variables.

The options described in the previous section are also applicable to the PLC under consideration. Since in this PLC we need an instruction at the end to ensure the execution of the last logic OR operation, this instruction may have, besides this objective, the purpose to act upon the output flip-flop as a function of the result. The instructions of a basic PLC, like the one being described, are either logic instructions or execution instructions.

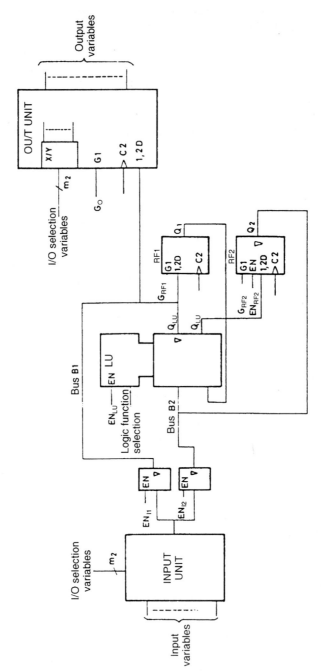

FIGURE 4.7 *Hardware of the operating unit of a basic PLC with conditional operating instructions.*

INTRODUCTION TO PLCS

TABLE 4.4

Instruction	Code (Octal)
AND	0
INVERTED AND	1
OR	2
INVERTED OR	3
EQUALITY	4

Logic instructions

Under this heading we have the logic AND function, the inverted logic AND function, the OR function and the inverted OR function. To each of them we assign a binary combination that can be represented by its octal or hexadecimal equivalent. The following codes are assumed:

    0 XXX   AND function
    1 XXX   Inverted AND function
    2 XXX   OR function
    3 XXX   Inverted OR function

The field XXX corresponds to an octal number representing the address of the input variable which will take part in the specified operation.

Each instruction is described below and the way it works is illustrated by means of a simple example.

LOGIC AND INSTRUCTION

This instruction is used to start a logic equation where the specified variable is used in a direct form. It also executes the logic AND function between the equation specified by the preceding logic instructions and the variable specified by the instruction.

EXAMPLE 4.1

Execute the logic product of input variables 101_8 and 407_8.

Solution

The instructions to be positioned in sequence in the program memory are:

    0 101
    0 407

INVERTED LOGIC AND INSTRUCTION

This instruction is also used to start a logic equation where the specified variable is used in the inverted form. It also executes the logic product between the equation specified by the preceding logic instructions and the inverse of the variable specified by the instruction.

EXAMPLE 4.2

Execute the logic product of the inverse of the input variable 210_8 and 314_8.

Solution
The instructions to be positioned in sequence in the program memory are:

      1 210
      0 314

From what has been said we conclude that a sequence of an AND and inverted AND functions constitutes a product of variables.

LOGIC OR INSTRUCTION
This instruction is used to separate the logic products and to execute their logic sum. The direct variable indicated in this instruction is part of the new product.

EXAMPLE 4.3

Implement a program which executes expression $\bar{a}\bar{b} + a\bar{c}$.

Solution
We assume variables a, b and c are assigned to the following positions:

      $a: 127_8$
      $b: 134_8$
      $c: 210_8$

The above expression is executed with the following instruction sequence:

      1 127
      1 134
      2 127
      1 210

INVERTED LOGIC OR INSTRUCTION
This instruction is used to separate the logic products and to execute their logic sum. The direct variable indicated in this instruction is part of the new product in its inverted form.

INTRODUCTION TO PLCS

EXAMPLE 4.4

Write a program to execute expression $a\bar{b}c + \bar{a}bd$.

Solution

We assume variables a, b, c and d are in the following input positions:

a: 127_8
b: 129_8
c: 210_8
d: 212_8

The above expression is executed with the following instruction sequence:

0 128
1 129
0 210
3 128
0 129
0 212

Executive instructions

These are instructions such as equality, change, set and reset. The following codes are assumed:

4 XXX Equality
5 XXX Change
6 XXX Set
7 XXX Reset

where symbol X is an octal number from 0 to 7 and field XXX represents the address of the data memory whose contents are to be used in the specified operation.

The mode of operation of these instructions is described below.

EQUALITY INSTRUCTION

This instruction equals the state of the specified output memory position to the logic value of the operation defined by the logic instructions that precede it.

EXAMPLE 4.5

Write a program which makes the output variable 354_8 equal to the logic value of the expression defined by the instructions of example 4.1.

Solution

After the instructions of example 4.1, we must add an equality instruction. Therefore, the required program is:

    0 101
    0 407
    4 354

From this example we conclude that the equality instruction is an execution instruction which separates the logic equations.

CHANGE INSTRUCTION

This instruction changes the state of the specified output variable only if the value of the equation defined by the preceding logic instructions is 1.

EXAMPLE 4.6

Write a program which changes the state of the output variable 127_8 if the state of position 214_8 or position 320_8 is 1.

Solution

The instructions to be programmed in sequence are:

    0 214
    2 320
    5 127

SET INSTRUCTION

This instruction sets the specified output variable if the logic value of the equation defined by the preceding logic instructions is 1. Its use is similar to that demonstrated in example 4.6.

RESET INSTRUCTION

This instruction resets the specified output variable if the value of the equation defined by the preceding logic instructions is a logic 1. Its use is also similar to that demonstrated in example 4.6.

Besides these instructions, basic PLCs may have others which will increase the flexibility of their use. One of them is, for example, 'Start the program'. When the PLC is turned on, its control unit remains idle, executing no action except reading the program memory. The reading of a 'Start program' instruction makes the control unit change to an active state and start the execution of the instructions stored in the program memory, beginning at that location.

4.4 Basic PLCs with improved characteristics

In sections 4.2 and 4.3 two basic PLCs with a minimum of hardware are analyzed. The addition of some functional blocks, made possible by the progress in microelectronics, results in PLCs with higher capacity and flexibility. In this section an example of such an additional functional block is described.

4.4.1 PLC with input and output memory units

The PLCs studied in sections 4.2 and 4.3 have the following characteristics:

1. Their input variables act directly upon the logic unit.
2. Their output variables are only stored in the output flip-flops.

In these PLCs each logic instruction reads an input variable and executes a logic operation simultaneously. The program follows the cycle shown in Figure 4.8, which combines input, processing and output actions sequentially.

The introduction of a random access memory (RAM) between the input unit and the logic unit, and another between the logic unit and the output unit, endows the PLC with the following advantages:

1. The memory of the input variables is a copy of their state at a given time.

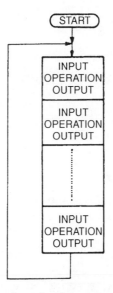

FIGURE 4.8 *Flowchart for the program of a basic PLC.*

2. The internal states of the state diagram of sequential systems may be also stored in the output variables' memory. In this way, the PLC can easily act like an asynchronous sequential system.

3. The program of the PLC may be executed by means of two distinct consecutive cycles which are periodically repeated, as represented in Figure 4.9:
    - an input and output cycle, where the input variables are stored into the input RAM and the information of the output variables is transferred from the output RAM to the output flip-flops;
    - a processing cycle where the instructions are executed.

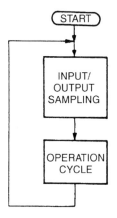

FIGURE 4.9 *Flowchart for the program of a PLC with input and output memory units.*

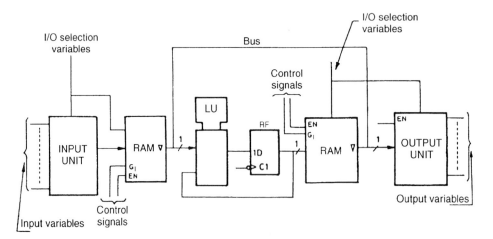

FIGURE 4.10 *Hardware of the operating unit of a PLC with input and output memory units.*

INTRODUCTION TO PLCS 143

During the input and output cycles every internal state flip-flop of the operating unit is set at the right state.

Figure 4.10 shows the operating unit of a PLC similar to that of Figure 4.3 including an input memory unit and an internal state and output memory unit.

4.5 Digital systems synthesis using PLCs

In the previous sections we have shown that a PLC can execute any logic function by an appropriate sequence of instructions. In this section, several practical examples are given to help consolidate the reader's knowledge. We use the PLC described in section 4.3 with an additional random access memory (RAM) which stores input, output and internal state variables. This PLC has the set of instructions listed in Table 4.5.

TABLE 4.5

Instruction	Code (Octal)
AND	0XXX
INVERTED AND	1XXX
OR	2XXX
INVERTED OR	3XXX
EQUALITY	4XXX
CHANGE	5XXX
SET	6XXX
RESET	7XXX

4.5.1 Combinational system synthesis

EXAMPLE 4.7

Write a program for the PLC, using the set of instructions listed in Table 4.5, to supervise the chemical process of example 1.1.

Solution

Figure 4.11 represents the block diagram of the system. The truth table obtained from the problem statement is shown again in Table 4.6.

The use of a PLC renders unnecessary the execution of any minimization process because it is not economically significant. Therefore, from the truth table the following expression is directly obtained:

$$f = T_1 \overline{T_2} \, \overline{T_3} + T_1 T_2 T_3$$

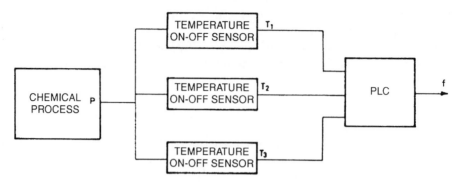

FIGURE 4.11 *Block diagram of the electronic supervision system of example 1.1, using a PLC.*

TABLE 4.6

T_3	T_2	T_1	f
0	0	0	0
0	0	1	1
0	1	0	X
0	1	1	0
1	0	0	X
1	0	1	X
1	1	0	X
1	1	1	1

TABLE 4.7

Program memory address	Instruction	Comment
000	0.021	AND T1
001	1.022	AND $\overline{T2}$
002	1.023	AND $\overline{T3}$
003	2.021	OR T1
004	0.022	AND T2
005	0.023	AND T3
006	4.200	Equals f to the equation value

INTRODUCTION TO PLCS

If it is assumed variables T_1, T_2 and T_3 are assigned to the input memory locations 21, 22 and 23 and f to location 200 of the output memory, the program starting from the location zero of the instruction memory is the one indicated in Table 4.7.

Therefore we conclude that a PLC can execute any logic equation of the variables present at its inputs.

4.5.2 Edge-characterized sequential control system synthesis

The synthesis of a sequential control system begins by establishing the operation specifications, which can be defined in terms of levels or transitions of the input variables.

Specifications in terms of input variable transitions allow us to carry out a systematic design. The PLC must be able to execute the different expressions of the transition capacity, which are:

$$C_{T1} = \sum_{h=1}^{h=n} x_h \updownarrow (X_\beta^h)$$

$$C_{T2} = \sum x_h \updownarrow (X'_\alpha)$$

$$C_{T3} = \sum x_h \updownarrow$$

$$C_{T4} = \sum X_a \updownarrow$$

These expressions specify the conditions capable of producing a change of the internal state of an asynchronous sequential system. Expression C_{T1} specifies that the change of internal state takes place when a generic variable x_h changes state and all the others remain at a certain state defined by X_β^h, where h varies from 1 to n.

Expression C_{T2} indicates that the change of internal state will take place when x_h changes and a certain number of input variables remain at a certain state.

Expression C_{T3} specifies the changes of the internal state due to changes of one input variable independently of the state of all the others.

Expression C_{T4} corresponds to the change of internal state due to the change of a vector of the input variables.

The reader wishing to study the specification of edge-characterized asynchronous sequential systems in greater detail is referred to [MAND 91], although the practical application of this method can be easily understood by means of several examples, as shown in the next section.

The transition capacity expressions given above include edges which must be converted to logic operations. Fortunately, the periodic sampling of the input variables allows the specification of level transitions or edges by logic products. In effect, an input variable changes state when its value at sampling instant t is different from the value at sampling instant $t-1$. As a consequence, level changes can be expressed algebraically:

$$x_i\uparrow \equiv x_{it}\overline{x_{it-1}} = 1$$
$$x_i\downarrow \equiv \overline{x_{it}}x_{it-1} = 1$$
$$x_i\updownarrow \equiv x_{it}\overline{x_{it-1}} + \overline{x_{it}}x_{it-1} = 1$$

These algebraic expressions for the level transitions or edges are combined with the state assignment method called 'one hot encoding (OHE)'. This method assigns one binary variable to every internal state. The value of this variable is a logic 1 only when the logic controller is in the corresponding internal state.

Programming transition capacity expressions

EXPRESSIONS C_{T1} AND C_{T2}

It is assumed that the transition from state E_i to state E_j (Figure 4.12) takes place when variable x_i changes from 0 to 1, provided the input vector formed by all the variables except x_i is X_a. It is also assumed the input variables are x_i, x_j and x_k and that $X_a = \bar{x}_j x_k$. Table 4.8 shows the program with a short comment beside each instruction. It is assumed the different variables are stored in the following positions of the data memory:

$E_i = 100_8 \quad x_{it-1} = 002_8$

$E_j = 101_8 \quad x_j = 003_8$

$x_{it} = 001_8 \quad x_k = 004_8$

The expression of C_T corresponding to Figure 4.12 is:

FIGURE 4.12 *Transition capacity example of expression C_{T2}.*

$$C_T = x_i\uparrow(\overline{x_j} \cdot x_k)$$

and expressing x_i by its algebraic equivalent results in:

$$C_T = x_{it}\overline{x_{it-1}}(\overline{x_j} \cdot x_k)$$

This equation is executed by means of instructions located in addresses $h+1$ to $h+4$ of the instruction memory (Table 4.8). Only if this equation is a logic one and the PLC is in state E_i, the instruction stored in address $h+5$ sets the data memory location corresponding to state E_j, and the instruction

TABLE 4.8

Program memory address (Octal)	Instruction	Comment
h	0.100	AND E_i
h+1	1.003	AND $\overline{x_j}$
h+2	0.004	AND x_k
h+3	0.001	AND x_{it} ⎫
h+4	1.002	AND $\overline{x_{it-1}}$ ⎭ $x_i\uparrow$
h+5	6.101	Set E_j
h+6	7.100	Reset E_i
h+7	0.001	⎫
h+10	4.002	⎭ $x_{it} \rightarrow x_{it-1}$

stored in address h + 6 resets the data memory location corresponding to state E_i. Instructions in addresses h + 7 and h + 10 store x_{it-1} to be compared with x_{it} during the next processing cycle.

The particular case where X^i_β is replaced by a vector X'_α of smaller dimension presents no difference in the way of establishing the PLC program to obtain C_T and consequently needs no further explanation (C_{T2} case).

EXPRESSION C_{T3}

This expression requires a smaller number of instructions than either C_{T1} or C_{T2}. Table 4.9 shows the appropriate program to ensure the transition from state E_i to E_j when x_i changes from 0 to 1 and the transition from state E_i to E_n when x_j changes from 1 to 0 (Figure 4.13). It is assumed that the different variables are kept in the following locations of the data memory:

$E_i = 200_8$ $x_{it} = 010_8$ $x_{it-1} = 011_8$

$E_j = 201_8$ $x_{jt} = 012_8$ $x_{jt-1} = 013_8$

$E_h = 202_8$

In this case, the transition capacity from E_i is:

$C_T = x_i\uparrow + x_j\downarrow$

$x_i\uparrow$ detection is done when instructions located in addresses P + 1 and P + 2 are executed. In the same way, instructions located in addresses P + 6 and P + 7 detect $x_j\downarrow$. Instructions located in addresses P + 12 and P + 13 transfer the state of x_{it} to x_{it-1} and those in P + 14 and P + 15 transfer the state of x_{jt} to x_{jt-1}.

TABLE 4.9

Program memory address (Octal)	Instruction	Comment
P	0.200	AND E_i
P+1	0.010	AND x_{it}
P+2	1.011	AND $\overline{x_{it-1}}$
P+3	6.201	Set E_j
P+4	7.200	Reset E_i
P+5	0.200	AND E_i
P+6	1.012	AND $\overline{x_{jt}}$
P+7	0.013	AND x_{jt-1}
P+10	6.202	Set E_h
P+11	7.200	Reset E_i
P+12	0.010	$x_{it} \to x_{it-1}$
P+13	4.011	
P+14	0.012	$x_{jt} \to x_{jt-1}$
P+15	4.013	

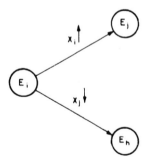

FIGURE 4.13 *Transition capacity example of expression C_{T3}.*

EXPRESSION C_{T4}

Another expression for transition capacity consists of the state change of an input vector. Table 4.10 shows the program which executes the transition from state E_i to E_j if vector X_a changes from 0 to 1. It is assumed the system has three input variables x_1, x_2 and x_3 and vector X_a is defined by the product $\overline{x_1}x_2\overline{x_3}$ (Figure 4.14).

INTRODUCTION TO PLCS

TABLE 4.10

Program memory address (Octal)	Instruction	Comment
P	0.200	AND E_i
P+1	1.100	AND $\overline{x_1}$ ⎫
P+2	0.101	AND x_2 ⎬ X_α
P+3	1.102	AND $\overline{x_3}$ ⎭
P+4	4.103	Equality $X_{\alpha t}$
P+5	0.103	AND $X_{\alpha t}$ ⎫
P+6	1.104	AND $\overline{X_{\alpha t-1}}$ ⎬ $X_\alpha \uparrow$
P+7	6.201	Reset E_j
P+10	7.200	Set E_i
P+11	0.103	⎫
P+12	4.104	⎬ $X_{\alpha t} \to X_{\alpha t-1}$

FIGURE 4.14 *Transition capacity example of expression C_{T4}.*

The following data memory locations are assigned to the variables:

$E_i = 200_8$ $x_3 = 102_8$
$E_j = 201_8$ $X_{at} = 103_8$
$x_1 = 100_8$ $X_{at-1} = 104_8$
$x_2 = 101_8$

The instructions located in addresses P+1 to P+3 execute vector X_{at}, and the change of state of this vector is detected by the instructions at addresses P+5 and P+6. The instructions at addresses P+11 and P+12 transfer the state of vector X_{at} to X_{at-1}.

Programming a transition graph

Finally, a complete simulation example of a transition graph or state diagram is presented. Figure 4.15 shows the proposed graph, which is identical to the graph of the cart control system of example 1.3.

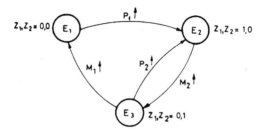

FIGURE 4.15 *Example of a transition graph or state diagram.*

TABLE 4.11

E_1	E_2	E_3	Z_1	Z_2
1	0	0	0	0
0	1	0	1	0
0	0	1	0	1

We start by assigning data memory locations to internal state variables. Using 'one hot encoding', output variables Z_1 and Z_2 are identical to internal states E_2 and E_3 respectively, as indicated in Table 4.11, and consequently they use the same data memory position. Only state E_1 requires a separate data memory location:

$E_1 = 200_8$

The data memory locations assigned to the input and output variables respectively are:

Variable	Location
M_1	001_8
M_2	002_8
P_1	003_8
P_2	004_8
Z_1	040_8
Z_2	041_8

The data memory locations that store the edge-active input variables during time $t-1$ are:

Variable	Location
M_{1t-1}	100_8
M_{2t-1}	110_8
P_{1t-1}	120_8
P_{2t-1}	130_8

TABLE 4.12

Program memory address (Octal)	Instruction	Comment	
0	1.300	AND $\overline{300}$	Initial State set-up
1	6.200	Set E_1	
2	6.300	Set 300	
3	0.200	AND E_1	
4	0.003	AND P_{1t}	$P_1 \uparrow$
5	1.120	AND $\overline{P_{1t-1}}$	
6	7.200	Reset E_1	
7	6.040	Set Z_1 and E_2	
10	0.040	AND E_2	
11	0.002	AND M_{2t}	$M_2 \uparrow$
12	1.110	AND $\overline{M_{2t-1}}$	
13	7.040	Reset Z_1 and E_2	
14	6.041	Set Z_2 and E_3	
15	0.041	AND E_3	
16	0.004	AND P_{2t}	$P_2 \uparrow$
17	1.130	AND $\overline{P_{2t-1}}$	
20	7.041	Reset Z_2 and E_3	
21	6.040	Set Z_1 and E_2	
22	0.041	AND E_3	
23	0.001	AND M_{1t}	$M_1 \uparrow$
24	1.100	AND $\overline{M_{1t-1}}$	
25	7.041	Reset Z_2 and E_3	
26	6.200	Set E_1	
27	0.001	$M_{1t} \rightarrow M_{1t-1}$	
30	4.100		
31	0.002	$M_{2t} \rightarrow M_{2t-1}$	
32	4.110		
33	0.003	$P_{1t} \rightarrow P_{1t-1}$	
34	4.120		
35	0.004	$P_{2t} \rightarrow P_{2t-1}$	
36	4.130		

To set the initial state, position 300 is used. At power-up, all the data memory is reset during an initialization cycle. The initial part of the program sets locations 100 and 300 of the data memory (state E_1) only if location 300 has a logic 0. In this fashion the PLC is set at the initial state of the transition graph when powered-up.

The program corresponding to this example is shown in Table 4.12. Each transition shown in the graph of Figure 4.15 is programmed in Table 4.12. Once this program is introduced in the PLC, it will behave in accordance with the above-mentioned transition graph.

In this example, the transition capacity expressions consist entirely of changes in the binary value of the input variables. Such changes are detected, as indicated in Table 4.12, by the logic product of the input variable during the current sampling cycle and the previous sampling cycle. For example, the change from 0 to 1 of the input variable P_1 is detected by the logic product of P_{1t} and $\overline{P_{1t-1}}$.

Next, we analyze another logic controller design using a PLC.

EXAMPLE 4.8

Write a program for the PLC with the set of instructions indicated in Table 4.5, which executes the control of the bar selection system described in example 3.3.

Solution

Figure 4.16 represents the state diagram obtained from the specifications of example 3.3.

We note that the output variable Z is the same as the internal state variable E_2. First, we assign data memory locations to input variables x_1 and x_2 and internal state variables E_1 and E_2 (Table 4.13). Similarly, position 010 is assigned to variable x_{2t-1}.

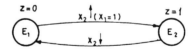

FIGURE 4.16 *Transition graph or state diagram of the bar selection system of example 1.4.*

TABLE 4.13

Variable	Location
x_1	001_8
x_2	002_8
E_1	020_8
E_2	021_8

INTRODUCTION TO PLCS

TABLE 4.14

Program memory address (Octal)	Instruction	Comment	
0	1.050	AND $\overline{50}$	Initial State set-up
1	6.020	Set E_1	
2	6.050	Set 50	
3	0.020	AND E_1	
4	0.002	AND x_{2t}	$x_2 \uparrow$
5	1.010	AND $\overline{x_{2t-1}}$	
6	0.001	AND x_1	
7	6.021	Set E_2	
10	7.020	Reset E_1	
11	0.021	AND E_2	
12	1.002	AND $\overline{x_{2t}}$	$x_2 \downarrow$
13	0.010	AND x_{2t-1}	
14	6.020	Set E_1	
15	7.021	Reset E_2	
16	0.002	$x_{2t} \to x_{2t-1}$	
17	4.010		

Table 4.14 shows the corresponding program which uses memory location 50 to set the initial state E_1. The comments on the right make the program self-explanatory.

4.5.3 Level-characterized sequential control system synthesis

The specification of an asynchronous sequential control system by the levels of the input variables makes a systematic implementation difficult and its execution with asynchronous logic controllers impractical [UNGE 57] [KOHA 70] [McCL 86]. Nevertheless, one of the methods to implement level-characterized asynchronous sequential control systems enjoyed some popularity among engineers owing to its being empirically feasible [MAND 84].

Such a method is shown in Figure 4.17. The asynchronous logic controller is formed by a set of asynchronous R–S flip-flops with inputs R and S

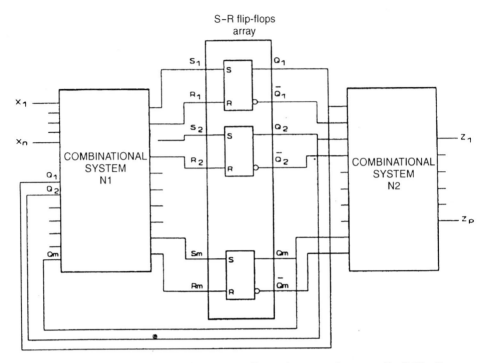

FIGURE 4.17 *Asynchronous logic controller using asynchronous R–S flip-flops.*

connected to the outputs of a combinational circuit. The inputs of this combinational circuit are the input variables x of the logic controller and the outputs Q of the flip-flops. The output variables of the logic controller can be obtained as a combination of variables Q or of these with inputs x.

The asynchronous R–S flip-flops can be easily simulated using a PLC. Figure 4.18a shows the logic diagram of an asynchronous R–S flip-flop constructed from two NOR gates. Figure 4.18b shows its relay contact equivalent circuit, which constitute a start–stop circuit with priority stop. M is the set variable and P is the reset variable and takes priority over M. Finally, Figure 4.18c shows the sequence of instructions of a PLC with the set of instructions described in section 4.5.1 (Table 4.5).

The system of Figure 4.17 has the drawback of lack of programmability and the possibility of hazards [FRIE 86] [MAND 84] which are eliminated using the PLC. The first disadvantage is overcome because the mode of operation can be changed by modifying the sequence of instructions, and the second because all input variables cannot change at the time instructions are executed.

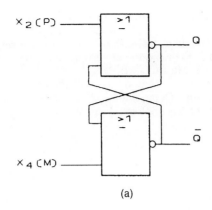

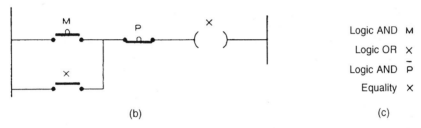

FIGURE 4.18 *Asynchronous R–S flip-flop with priority reset.* (a) *NOR gate implementation.* (b) *Relay implementation.* (c) *Implementation using a PLC with the instruction set of Table 4.5.*

On the other hand, the use of asynchronous R–S flip-flops, especially the version using relay contacts, makes the design of electronic control systems much more intuitive than with the characterization by edges.

Examples 5.3 and 5.4 of chapter 5 analyze electronic control systems implemented with a PLC using characterization by levels of the input variables.

Bibliography

[AN 76]   H.K. An et al. *Boolean processor hardware description*, ITT Technical Report No. STL 1256, 1976.
[FRIE 86] A.D. Friedman, *Fundamentals of Logic Design and Switching Theory*. Computer Science Press, 1986.
[KOHA 70] Z. Kohavi, *Switching and Finite Automata Theory*. McGraw-Hill, 1970.
[MAND 84] E. Mandado, *Sistemas electrónicos digitales*, 5th edition, section 6.2.2. Marcombo, 1984.

[MAND 91]   E. Mandado, *Sistemas electrónicos digitales*, 7th edition. Marcombo, 1991.
[McCL 86]   E.J. McCluskey, *Logic Design Principles*. Prentice Hall, 1986.
[MOTO 77]   Motorola Semiconductor Products, Inc., *Industrial Control Unit Handbook*, MC145000B, 1977.
[UNGE 57]   S.H. Unger, 'A study of asynchronous logical feedback networks', PhD dissertation, Department of Electrical Engineering, Massachusetts Institute of Technology, June 1957.

CHAPTER 5
PLC programming languages

5.1 Introduction

Programming a PLC consists in establishing an ordered sequence of instructions (available in the equipment) which solve a certain algorithm corresponding to a control task. The sequence which establishes the relation between the different logic variables constitutes the program.

In the previous chapter it is shown that each PLC has a certain machine code language according to its hardware architecture. It is demonstrated in section 4.5 that it is possible to program a PLC by directly establishing a set of instructions in machine code language. However, machine code language is far different from the language used by specialists to specify the operation of a control system. Therefore, PLC manufacturers use a variety of programming languages in their equipment, namely:

- Instruction list
- Relay or ladder diagram
- Function diagram
- Grafcet

All these languages make the programming task easier. The choice of one or the other depends on the experience and knowledge of the user (in digital electronics, computing, implementation of control systems with relays, etc.) and the way the control problem is specified.

There is not a unique description for each language although some standards have been developed. Instead, each manufacturer uses a particular designation for the different instructions and also a particular configuration to represent the different input and output (internal or external) variables.

In this chapter we use a set of instructions and variables representation corresponding to a general PLC. In the next chapter, several commercial PLCs are analyzed and used to solve various practical examples.

5.2 Instruction list

The instruction list language consists of a set of symbolic codes each corresponding to an instruction in machine code language. Since PLC machine languages differ between manufacturers, the instruction list languages too are different.

Because a programming language using symbolic codes is very close to machine code language, it is especially suitable for users familiar with digital electronics and computing. On the other hand, this language is the only one that can be used with simple programming units which can display only a few program lines simultaneously.

In the following sections, the way of representing PLC variables and instructions is analyzed.

5.2.1 Variable identification

According to the type of variable, variable identification is performed as follows:

1. **Input variables** Xn. Symbol X indicates a binary input variable and it is assigned with number n corresponding to its address or location in the input connector.
2. **External output variables** Yn. Symbol Y indicates a binary output variable and it is assigned with number n corresponding to its address or location in the output connector.
3. **Internal output variables** IRn. Symbol IR represents an internal binary variable (memory element) and n is the corresponding order number.

5.2.2 Instructions

It is assumed that the PLC has three different kinds of instruction, analyzed next.

Selection, input and output or operation instructions

These instructions execute one of the following actions:

1. Select a certain variable to be used as an operand, an input variable or an output variable.
2. Execute an input or an output.
3. Execute a certain operation with a given variable.

It is assumed that the PLC has within this group the following ten instructions.

PLC PROGRAMMING LANGUAGES   159

STR
This instruction is used to select the first variable in a sequence of instructions. For example, instruction STR X0 selects input variable X0, while STR Y6 selects external output variable Y6 and STR IR2 selects internal output variable IR2.

STR NOT
This instruction is used to select the first inverted variable in a sequence of instructions. For example, instruction STR NOT X12 selects input variable X12 and inverts it ($\overline{X\,12}$); instruction STR NOT Y10 selects external output variable Y10 and inverts it ($\overline{Y\,10}$) and instruction STR NOT IR9 selects internal output variable IR9 and inverts it ($\overline{IR\,9}$).

OUT
This instruction acts upon the corresponding output variable (external or internal). For example, the sequence:

    STR X0
    OUT Y0

selects input variable X0 [STR X0] and transfers its value to the external output variable Y0.

OUT NOT
This instruction acts upon the corresponding inverted output variable (external or internal). For example, the sequence:

    STR X0
    OUT NOT Y0

selects input variable X0 and equalizes the external output variable Y0 to $\overline{X0}$ (inverse of the input variable X0).

OR
This instruction executes the logic OR function between one variable or combination of variables and the variable specified in it. For example, the sequence of instructions:

    STR Y5
    OR X3
    OR IR7
    OUT Y0

executes the following actions:

- selects the external output variable Y5 [STR Y5];
- executes the logic OR between Y5 and X3 [OR X3];

- executes the logic OR of IR7 and the result of the logic OR between Y5 and X3 [OR IR7];
- makes the external output variable Y0 equal to the result of the previous instruction in the sequence [OUT Y0].

The result of this instruction sequence is:

Y0 = Y5 + X3 + IR7

OR NOT

This instruction executes the logic OR function between one variable or combination of variables and the inverse of the variable specified in it. As an example, the reader may analyze the following sequence of instructions:

 STR IR13
 OR NOT X10
 OR NOT X14
 OUT IR15

and verify that it executes the function IR15 = IR13 + $\overline{X10}$ + $\overline{X14}$.

AND

This instruction executes the logic AND function between one variable or combination of variables and the variable specified in it. For example, the sequence:

 STR NOT X0
 AND X1
 AND IR7
 AND Y3
 OUT Y10

executes the following actions:

- selects the input variable X0 and inverts it [STR NOT X0];
- executes the logic AND between $\overline{X0}$ and X1 [AND X1];
- executes the logic AND between IR7 and the result of the logic AND between $\overline{X0}$ and X1 [AND IR7];
- executes the logic AND between output variable Y3 and the logic product $\overline{X0}$ · X1 · IR7 [AND Y3];
- makes the external output variable Y10 equal to the result of the previous instruction sequence [OUT Y10].

The result of this instruction sequence is:

Y10 = $\overline{X0}$ · X1 · IR7 · Y3

N.B. Instruction AND after an OR executes the logic AND of one variable with the logic OR of the previous instructions. For example, the instruction

PLC PROGRAMMING LANGUAGES

sequence:

    STR X5
    OR X3
    AND Y5
    OUT Y3

executes the following function:

$Y3 = (X5 + X3) \cdot Y5$

AND NOT

This instruction executes the logic AND function between one variable or combination of variables and the inverse of the variable specified in it. As an example, the reader may analyze the following sequence of instructions:

    STR Y6
    AND NOT X3
    AND NOT IR9
    AND NOT X9
    OUT IR14

and verify that it executes the function $IR14 = Y6 \cdot \overline{X3} \cdot \overline{IR9} \cdot \overline{X9}$.

OR STR

This instruction executes logic OR between the two previous sequences initiated by STR or STR NOT. For example the instruction sequence:

    STR X7
    OR X9
    AND NOT Y5
    STR NOT IR3
    AND X6
    OR NOT Y6
    OR STR
    OUT Y8

executes the following actions:

- selects the input variable X7 [STR X7];
- executes the logic OR between X7 and X9 [OR X9];
- executes the logic AND between $\overline{Y5}$ and X7 + X9 [AND NOT Y5];
- initiates another sequence selecting variable IR3 inverted [STR NOT IR3];
- executes the logic AND between $\overline{IR3}$ and X6 [AND X6];
- executes the logic OR between $\overline{Y6}$ and $\overline{IR3} \cdot X6$ [OR NOT Y6];

- executes the logic OR between the two expressions $(X7+X9) \cdot \overline{Y5}$, and $\overline{IR\,3} + X6 \cdot \overline{Y6}$ [OR STR];
- equalize the external output variable Y8 to the previous expression [OUT Y8].

This instruction sequence executes the function:

$$Y8 = [(X7+X9) \cdot \overline{Y5}] + (\overline{IR\,3} \cdot X6 + \overline{Y6})$$

AND STR

This instruction executes the logic AND function between the two previous sequences initiated by STR or STR NOT. As an example, the reader may analyze the following instruction sequence:

```
STR X0
AND NOT X1
STR X2
AND X3
OR NOT Y0
AND STR
OUT Y1
```

and verify that it executes the function $Y1 = X0 \cdot \overline{X1} \cdot (X2 \cdot X3 + \overline{Y0})$.

Timing and counting instructions

Timing and counting instructions generate variables whose switching-on, duration or switching-off is a function of time or of the number of pulses applied to an input variable. It goes without saying that a PLC that implements these instructions will need timers and counters amongst its hardware. It is also assumed that the PLC has the following two instructions: TMR and CTR.

TMR

Instruction TMR uses two variables to execute the timing function, namely the reset variable Xi and the timing variable Xj. The output of the timer can be an internal or external output variable.

The programming of the timer needs four instructions in sequence:

1. One instruction to select the reset variable Xi.
2. One instruction to select the timing variable Xj.
3. The instruction TMR n which chooses the timer (n). TMR n initiates the timing if Xi is a 1 (there is no reset) and Xj becomes a 1 (the timer input variable is switched on).
4. One program memory location to store the value of the preselected time.

PLC PROGRAMMING LANGUAGES 163

An example of the generation of a timing variable Y0 is as follows:

    STR X1
    STR X0
    TMR 0
    10
    OUT Y0

Figure 5.1 shows the timing diagram of input signals X0 and X1 and output Y0. Y0 becomes a 1 only if X1 is a 1 [STR X1], X0 becomes a 1 [STR X0] and 10 time units have elapsed (for example, tenths of a second).

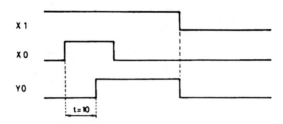

FIGURE 5.1 *Timing diagram of a timer.*

Instruction TMR may be used to delay a variable switching on. Hence, the same variable is used as a reset and timing variable. As an example, the sequence:

    STR X5
    STR X5
    TMR 2
    10
    OUT Y5

delays the output variable Y5 switching on ten time units from the instant that input X5 is switched on, as indicated in the timing diagram of Figure 5.2.

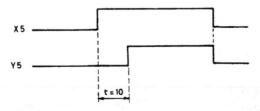

FIGURE 5.2 *Timing diagram of a delay to switching on.*

Similarly, TMR may be used to delay the switching on of a variable if it is selected inverted as a reset and timing variable. For example, the sequence:

>   STR NOT X4
>   STR NOT X4
>   TMR 6
>   10
>   OUT NOT Y9

delays the output variable Y9 switching off ten time units from the instant that input X4 is switched off, as indicated in the timing diagram of Figure 5.3.

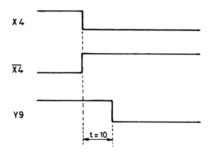

FIGURE 5.3 *Timing diagram of a delay to switching off.*

CTR

Instruction CTR executes the counting function, for which two or three variables may be used. In the first case, it acts as an 'up counter' where the first variable is the reset variable and the second is the counting one. In the second case, it acts as an up/down counter where the first and third variables have the same function as in the previous case and the second selects the counting mode, up or down, according to whether it is a 0 or a 1, respectively.

Therefore the programming of a counter needs four or five instructions in sequence operating in a similar manner to the instruction TMR discussed above. The following examples clarify these issues.

A sequence of instructions executing an up counter is the following:

>   STR X1
>   STR X0
>   CTR 3
>   10
>   OUT Y2

Variable Y2 becomes a 1 after the application of ten pulses to input X0 provided X1 is kept at level 1, and returns to 0 simultaneously with X1. Figure 5.4 shows the corresponding timing diagram.

PLC PROGRAMMING LANGUAGES   165

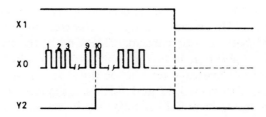

FIGURE 5.4 *Timing diagram of an up counter.*

A sequence of instructions executing an up/down counter is the following:

    STR X2
    STR X1
    STR X0
    CTR 4
    10
    OUT Y3

The timing diagram is depicted in Figure 5.5.

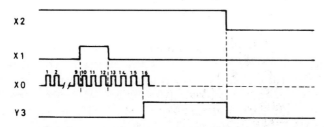

FIGURE 5.5 *Timing diagram of an up/down counter.*

Control instructions

Control instructions influence the execution of other instructions. Although a PLC can do without them, as shown in the basic PLC studied in chapter 4, their use simplifies the programming task.

There are many different ways of implementing control instructions and the PLC designer may choose those that seem most convenient for the job in hand.

We assume the PLC being studied has the following two pairs of control instructions.

JMP–JME
These instructions cause the instructions located between them to be executed or not depending on the result of the logic operation immediately

before JMP. If the result of this operation is 1, the instructions located between JMP and JME execute normally. Hence, the outputs (internal or external) selected between instructions JMP and JME are updated. If the result of the logic operation is 0, the instructions of the program between JMP and JME are not executed and hence the state of any output (external or internal) selected between JMP and JME is not modified.

As an example, let us examine the following instruction sequence:

    STR X0
    OUT Y0
    JMP
    STR Y0
    AND X0
    OUT Y1
    STR X1
    OUT IR2
    JME

First, the input variable X0 is selected [STR X0] and then Y0 equals X0 [OUT Y0]. Next, JMP is executed, and depending on whether the value of variable Y0, selected in the previous instruction, is 0 or 1, the PLC executes or not the actions corresponding to the instructions situated between JMP and JME.

One way to implement this is to include in the operating unit a disabling flip-flop which will turn off a section of the program. This flip-flop is set by instruction JMP if the result of the operation which precedes it is a 0, and does not modify its state if it is a 1. Instruction JME resets the flip-flop. The reader may verify that instructions JMP and JME act in the same way as CONDIS and UNCENA of the basic PLC studied in section 4.2.

IL-ILC

This pair of instructions causes all the outputs (external or internal) selected between them to be normally updated or to be reset, depending on whether the result of the logic operation immediately before IL is a 1 or a 0, respectively. As an example, the following sequence of instructions is analyzed:

    STR X7
    OUT Y10
    IL
    STR X9
    OUT Y9
    STR X15
    OUT Y15
    ILC

PLC PROGRAMMING LANGUAGES 167

First, the input variable X7 is selected [STR X7] and Y10 = X7 is executed [OUT Y10]. Next, IL is executed and, depending on whether the value of the variable Y0 selected in the previous instruction is 1 or 0, the four instructions preceding ILC are executed or not. If they are not executed then Y9 and Y15 are reset.

It is left as an exercise to the reader to configure the hardware needed to carry out these actions.

5.3 Relay or ladder diagrams

In this programming language, the task the PLC has to perform is represented graphically by a ladder or contact diagram. This programming language is specially designed:

- to simplify the replacement of a relay-based control system by a PLC;
- to simplify PLC programming for control engineers familiar with the design of relay-based control systems.

To program in this language we require a programming unit with a semi-graphical or graphical screen in order to display the ladder diagram.

The representation of logic variables and instructions in this language are looked at next.

5.3.1 Variable identification

Binary variables are represented by means of contacts identified by one alphabetical and one numerical symbol obtained from the instruction list described in section 5.2.1. The symbol used to represent a contact may comply with different standards, as shown in Figure 5.6. For example, input variable X3 is represented according to the standards DIN 40713-16 in Figure 5.7a and NEMA in Figure 5.7b, respectively.

The contacts of Figures 5.6 and 5.7 are normally open and represent direct variables. The inverse variables are represented by normally closed contacts as indicated in Figure 5.8 for the internal output or internal state variable $\overline{\text{IR }2}$.

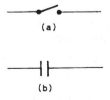

FIGURE 5.6 *Standard graphic symbol of an input, external output or internal output variables, using the ladder diagram.* (a) *DIN 40713–16 standard.* (b) *NEMA standard.*

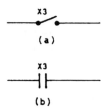

FIGURE 5.7 *Standard graphic symbol of input variable X3.*

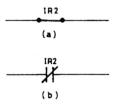

FIGURE 5.8 *Standard graphic symbol of internal variabe $\overline{IR2}$ used as an input variable.*

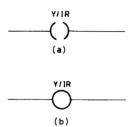

FIGURE 5.9 *Standard graphic symbol of an internal or external output variable.*

The function of the external or internal output generated by a combination of input variables (X), internal output (IR) or external output (Y) is represented in both standards as in Figure 5.9.

5.3.2 Logic sequences

The different logic functions can be represented in the relay or ladder diagram language. Next, we represent the ladder diagram and the instruction list corresponding to each function to demonstrate their equivalence.

FUNCTION OR INSTRUCTION OF A DIRECT INPUT VARIABLE
This function is represented by a normally open contact which, in general, activates an output variable, as indicated in Figure 5.10.

PLC PROGRAMMING LANGUAGES 169

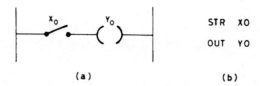

FIGURE 5.10 *Selection instruction of a normally open contact.* (a) *Ladder diagram.* (b) *Instruction list.*

FUNCTION OR SELECTION INSTRUCTION OF AN INVERTED INPUT VARIABLE
This function is represented by a normally closed contact which, in general, activates an output variable, as indicated in Figure 5.11.

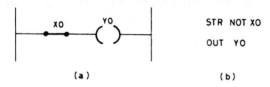

FIGURE 5.11 *Selection instruction of a normally closed contact.* (a) *Ladder diagram.* (b) *Instruction list.*

LOGIC OR FUNCTION OR INSTRUCTION
This function is represented by a set of parallel contacts which may be normally open (Figure 5.12), normally closed (Figure 5.13) or a combination of both.

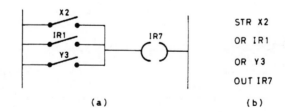

FIGURE 5.12 *Logic OR instruction.* (a) *Ladder diagram.* (b) *Instruction list.*

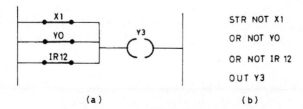

FIGURE 5.13 *Logic OR instruction.* (a) *Ladder diagram.* (b) *Instruction list.*

LOGIC AND FUNCTION OR INSTRUCTION

This function is represented by a series of contacts which may be normally open (Figure 5.14), normally closed (Figure 5.15) or a combination of both.

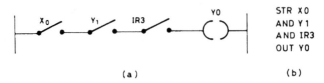

          (a)   (b)

FIGURE 5.14 *Logic AND instruction.* (a) *Ladder diagram.* (b) *Instruction list.*

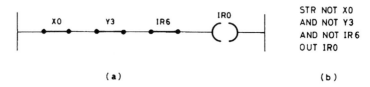

          (a)   (b)

FIGURE 5.15 *Logic AND instruction.* (a) *Ladder diagram.* (b) *Instruction list.*

LOGIC OR FUNCTION OR INSTRUCTION OF AND FUNCTIONS

This function is represented by a parallel combination of serial contacts as indicated in Figure 5.16.

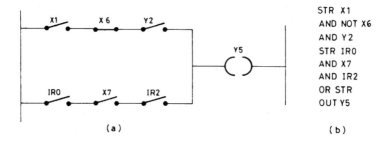

          (a)   (b)

FIGURE 5.16 *Logic OR of AND functions.* (a) *Ladder diagram.* (b) *Instruction list.*

LOGIC AND FUNCTION OF OR FUNCTIONS

This function is represented by a series of parallel contacts, as indicated in Figure 5.17. Combining this function with the previous one, more complex functions are obtained such as in Figure 5.18.

TIMING FUNCTION (TMR)

This function is represented by a functional block as in Figure 5.19. X1 is the timing variable, X0 executes the initialization or reset of the timer and Y0 is the output related to the timer.

PLC PROGRAMMING LANGUAGES   171

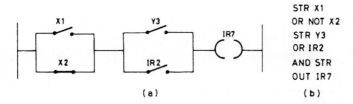

FIGURE 5.17 *Logic AND of OR functions.* (a) *Ladder diagram.* (b) *Instruction list.*

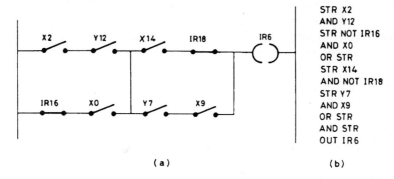

FIGURE 5.18 *Complex function combining logic OR of AND functions and logic AND of OR functions.*

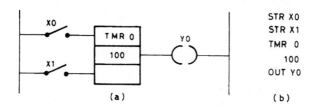

FIGURE 5.19 *Timing function.*

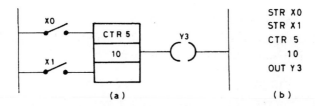

FIGURE 5.20 *Up-counting function.*

COUNTING FUNCTION (CTR)

This function has two variants, corresponding to an up or an up/down counter, each with a different functional block.

Figure 5.20 represents an up counter where X1 is the pulse-counting input and X0 the initialization or reset input.

Figure 5.21 represents an up/down counter where X2 is the pulse-counting input, X0 is the reset input and X1 is the direction of counting selection input (level 0 selects up-counting and level 1 down-counting).

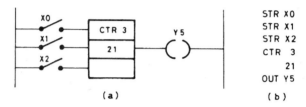

FIGURE 5.21 *Up/down counting function.*

5.4 Function diagram

A function diagram constitutes a symbolic language where the different combinations between variables are represented by the standard logic symbols described in appendix 1.

This programming language is particularly recommended to users familiar with digital electronics and, like the contact or ladder diagram, it needs a programming unit with a screen to display the diagram.

The representation of variables and instructions in the function diagram language is described below.

5.4.1 Variable identification

The same notation as in section 5.2.1 is used.

5.4.2 Logic operations

DIRECT INPUT VARIABLE SELECTION FUNCTION
This function is represented by the standard symbol of Figure 5.22a, which corresponds to the instruction list of Figure 5.22b.

INVERTED INPUT VARIABLE SELECTION FUNCTION
This function is represented by the standard symbol of Figure 5.23a, which corresponds to that of Figure 5.22a with the inversion symbol.

PLC PROGRAMMING LANGUAGES

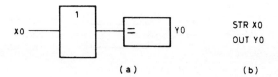

```
STR X0
OUT Y0
```

FIGURE 5.22 *Direct input variable selection function using function diagram language.*

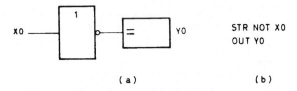

```
STR NOT X0
OUT Y0
```

FIGURE 5.23 *Inverted input variable selection function using function diagram language.*

LOGIC OR FUNCTION

This function may be implemented with either inverted or direct variables (Figure 5.24) or a combination of both (Figure 5.25).

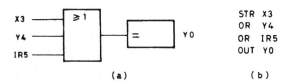

```
STR X3
OR  Y4
OR  IR5
OUT Y0
```

FIGURE 5.24 *Logic OR function using function diagram language.*

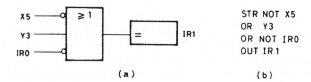

```
STR NOT X5
OR  Y3
OR  NOT IR0
OUT IR1
```

FIGURE 5.25 *Logic OR function using function diagram language.*

LOGIC AND FUNCTION

This function may be implemented with either inverted or direct variables (Figure 5.26) or a combination of both (Figure 5.27).

LOGIC OR FUNCTION OF AND FUNCTIONS

This function is implemented with a combination of AND and OR gates as shown in Figure 5.28.

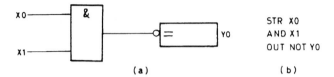

FIGURE 5.26 *Logic AND function using function diagram language.*

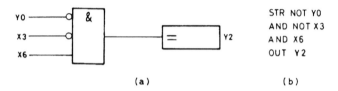

FIGURE 5.27 *Logic AND function using function diagram language.*

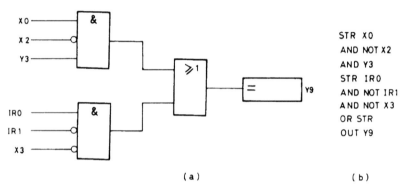

FIGURE 5.28 *Logic OR function of logic AND functions using function diagram language.*

LOGIC AND FUNCTION OF OR FUNCTIONS

This function is implemented with a combination of OR and AND gates as shown in Figure 5.29. By combining AND and OR gates, any complex function may be implemented, such as that shown in Figure 5.30.

TIMING FUNCTION

The timing function is represented by a functional block (Figure 5.31) with input variables X1 and X0. Variable X1 is the timing variable and X0 executes the initialization (reset) of the timer. The output variable is Y0. The operation mode is as described in section 5.2.2.

COUNTING FUNCTION

The counting function is also represented by a functional block with input and output variables. As described in the two languages studied earlier, CTR

PLC PROGRAMMING LANGUAGES

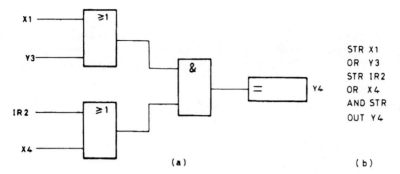

FIGURE 5.29 *Logic AND function of logic OR functions using function diagram language.*

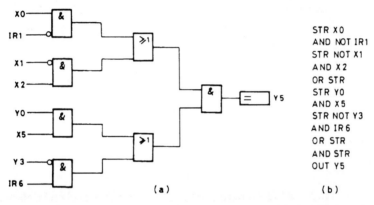

FIGURE 5.30 *Complex function combining logic AND and logic OR functions using function diagram language.*

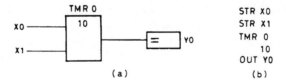

FIGURE 5.31 *Timing function using function diagram language.*

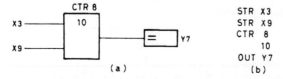

FIGURE 5.32 *Up-counting function using function diagram language.*

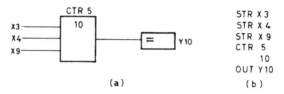

FIGURE 5.33 *Up/down-counting function using function diagram language.*

has two variants as shown in Figures 5.32 and 5.33. The mode of operation is as described in section 5.2.2.

5.5 Grafcet

As seen in sections 1.3 and 4.5.3, the flowchart or evolution graph is the best way of specifying a sequential control system. In section 4.5.3 it is also shown that the flowchart can be systematically converted into a PLC in machine code, instruction list, contact diagram or function diagram language. This has spurred interest in the standardization of the graphic representation of sequential control systems in order to enable the control engineers to design their own programs even though they may have limited knowledge of PLCs. In turn, the use of standard graphic specification methods has encouraged manufacturers to design programs capable of translating these graphic symbols into an instruction sequence that is stored in the PLC memory.

5.5.1 Definitions and fundamental symbols of Grafcet

Grafcet has been developed by a committee for the standardization of the specification of logic control systems appointed by the French Association for Technical and Business Cybernetics (AFCET).

Grafcet is based on the stage and receptivity concepts created by P. Girard [GIRA 73] which have been used in the specification by levels of logic control systems, and can also be used with the state and transition capacity concepts used in characterization by edges [MAND 91].

The standard symbols and the combinations used to represent the flowchart of a logic controller are described next.

The internal state represented by a circle (Figure 5.34a) is converted into a square with a number corresponding to the state (Figure 5.34b). If it is the initial state, it is represented by a double square (Figure 5.35) with a 0 inside.

The actions to be performed at a given internal state are described inside a rectangle linked to the corresponding state as in Figure 5.36.

A transition between two states is represented by a line with a stroke and labelled with the transition capacity C_T which should be 1 if the transition has to take place (Figure 5.37).

PLC PROGRAMMING LANGUAGES

FIGURE 5.34 *Graphic symbol of an internal state.* (a) *Symbol used in the flow diagram.* (b) *Symbol used in the Grafcet language.*

FIGURE 5.35 *Initial internal state graphic symbol using Grafcet.*

FIGURE 5.36 *Graphic symbol used to specify an action to be performed at a given internal state.*

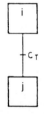

FIGURE 5.37 *Transition between two internal states.*

Normally, the state squares are aligned vertically and no arrow is used when the transition is downwards (Figure 5.37). If the transition takes place upwards, an arrow is included. When more than one transition is possible, the representation shown in Figure 5.38 is used; and when a state can be reached from different states (one at a time), one of the representations of Figure 5.39 is used.

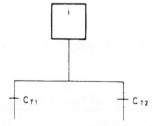

FIGURE 5.38 *Transition from one state to two different internal states.*

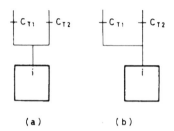

FIGURE 5.39 *Transition from two states to another different state.*

5.5.2 Application example

As an example, the Grafcet representation of the flowchart of Figure 5.40a, which was used as an example in section 4.5.3, is executed. Figure 5.40b represents the standard diagram in Grafcet. This is self-explanatory taking into account the rules presented above.

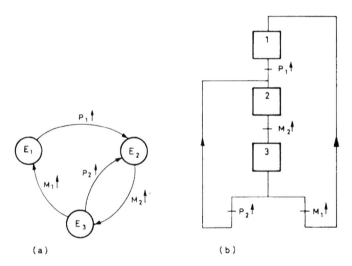

FIGURE 5.40 (a) *Example of flow or state diagram.* (b) *Example of standard flow diagram using Grafcet.*

5.6 Practical examples of digital system implementation using PLCs

5.6.1 Implementation of combinational systems using a PLC

In designing combinational systems using PLCs it is not necessary to obtain the minimized equations of the output variables, contrary to the situation

PLC PROGRAMMING LANGUAGES 179

when designing a logic controller with PLDs. The logic table is not necessary either, since it is straightforward to deduce from the problem statement in what conditions each output variable must be switched on.

Next, we use as an example the system implemented using PLDs in example 3.1, so that the reader may compare both methods of implementation.

EXAMPLE 5.1

Design a PLC program in the instruction list, ladder diagram and function diagram languages, to supervise the system of example 3.1.

Solution

From the criteria given above and the statement of the problem, we can deduce that the output variable PP, controlling the pump, should be a 1 if the tank level drops below a certain value (LSL = 1), or the tank temperature drops below a certain value (TSL = 1), or the pressure drop in filter F increases above a certain value (DPSH = 1), or the pressure in the fuel-oil collector decreases below a certain value (PSL = 1) or if the two flame detectors BS1 and BS2 indicate at the same time (BS1 = BS2 = 1) that the burners are off. Consequently, we have:

$$PP = LSL + TSL + DPSH + PSL + BS1 \cdot BS2$$

TABLE 5.1

External variable	PLC of section 5.2 variable assignment
LSL	X0
TSL	X1
DPSH	X2
PSL	X3
PSH	X4
BS1	X5
BS2	X6
PP	Y1
XV1	Y2
XV2	Y3
XV3	Y4
GL	Y5
RL	Y6

TABLE 5.2

Instruction list		Comment
STR	X0	
OR	X1	
OR	X2	
OR	X3	PP variable generation
STR	X5	
AND	X6	
OR STR		
OUT	Y1	
STR NOT	X5	XV1 variable generation
OUT	Y2	
STR NOT	X6	XV2 variable generation
OUT	Y3	
STR	Y1	
OR	X4	XV3 variable generation
OUT	Y4	
STR NOT	Y1	GL variable generation
OUT	Y5	
STR	Y1	RL variable generation
OUT	Y6	

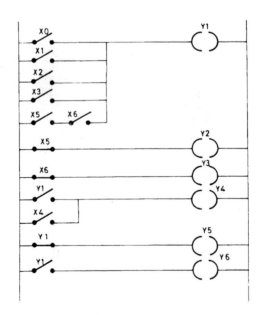

FIGURE 5.41 *Program corresponding to the supervision system of example 5.1 using the ladder diagram language.*

PLC PROGRAMMING LANGUAGES

We invite the reader to obtain, from the statement of the problem, the equations:

$$XV1 = \overline{BS1}; \quad XV2 = \overline{BS2}; \quad XV3 = PP + PSH; \quad GL = \overline{PP}; \quad RL = PP$$

Before writing the program, it is necessary to assign the input and output variables. This is shown in Table 5.1.

The instruction list program is shown in Table 5.2 with explanatory comments on the right.

Figures 5.41 and 5.42 show the program in ladder diagram and in function diagram, respectively. From sections 5.3 and 5.4, the reader should have no problem in arriving at them, and we recommend that readers attempt this exercise on their own and then compare their results with the diagrams given.

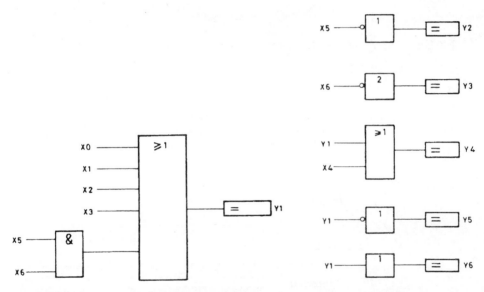

FIGURE 5.42 *Program corresponding to the supervision system of example 5.1 using the function diagram language.*

5.6.2 Implementation of edge-characterized sequential control systems using a PLC

In this section we use the method described in section 4.5.2, which constitutes a systematic and efficient design method. The solution of the problem of example 3.4 using a PLC is presented.

EXAMPLE 5.2

Write a PLC program in instruction list and Grafcet languages to control the bar selection system of example 1.4.

Solution

Figure 5.43 represents the flowchart that was obtained from the problem statement in example 3.3. From this flow diagram and applying the method described in section 4.5.2, the program can be written systematically.

The assignment of input and output variables is shown in Table 5.3.

One hot encoding is done, assigning to the internal states E_1 and E_2 the states of the internal variables, as shown in Table 5.4.

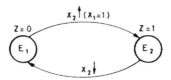

FIGURE 5.43 *Flow or state diagram of the bar selection system of example 1.4.*

TABLE 5.3

External variable	PLC of section 5.2 variable assignment
X_1	X1
X_2	X2
Z	Y0

TABLE 5.4

Internal state	PLC of section 5.2 internal state variable	
	IR0	IR1
E_1	1	0
E_2	0	1

To detect the rising and falling edges of X2, internal variable IR2 is used. This variable stores variable $X2_{t-1}$.

To set the initial state, variable IR3 is used. This variable resets at power-up like all the other IR.

Table 5.5 shows the program in instruction list language, which is self-explanatory with the adjoining comments. The conditional actions are executed by the instruction pair JMP–JME.

Figure 5.44 represents the program in Grafcet. This program was obtained almost immediately from the flow or state graph of Figure 5.43.

PLC PROGRAMMING LANGUAGES

TABLE 5.5

```
Instruction list                    Comment

STR NOT    IR3      ⎫
JMP                 ⎪
OUT        IR0      ⎬  Initial state set up
OUT        IR3      ⎪
JME                 ⎭

STR        IR0      ⎫
AND        X2       ⎪
AND NOT    IR2      ⎪  } x₂ ↑ (x₁ = 1)
AND        X1       ⎪
JMP                 ⎬                          E₁ → E₂
OUT NOT    IR0      ⎪
OUT        IR1      ⎪
OUT        Y0       ⎪
JME                 ⎭

STR        IR1      ⎫
AND NOT    X2       ⎪  } x₂ ↓
AND        IR2      ⎪
JMP                 ⎬                          E₂ → E₃
OUT NOT    IR1      ⎪
OUT        IR0      ⎪
OUT NOT    Y0       ⎪
JME                 ⎭

STR        X2       ⎫
OUT        IR2      ⎬  x₂ₜ → x₂ₜ₋₁
```

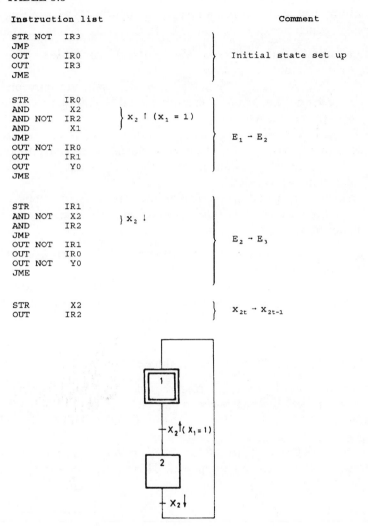

FIGURE 5.44 *Program corresponding to example 5.3 using Grafcet language.*

5.6.3 Implementation of level-characterized sequential control systems using a PLC

This form of implementation is based on the simulation of asynchronous S–R flip-flops in a PLC, as shown in section 4.5.3. This method is not entirely systematic and is more trial-and-error based. It starts from the problem statement, and the program is modified in a controlled way as new conditions are introduced.

This lack of a systematic procedure can be justified by the existence of several alternative ways to implement a sequential system using S–R flip-flops, two of which are:

1. **Design orientated towards the output variables**. The following general rules should be applied:
   - if, at a given time, an output variable depends only on the input variables at that time, a combinational circuit with the appropriate logic equation is assigned;
   - if, at a given time, an output variable depends on the value or the sequence of values of certain input variables at previous instants, an S–R flip-flop is assigned. This flip-flop is set or reset if certain conditions are verified;
   - additional internal states are created by S–R flip-flops to store the sequences which eventually cause an output S–R flip-flop to be switched on.

2. **Design orientated towards the internal state variables**. This design method consists of the following stages:
   - from the specifications we define the number of internal state variables necessary to store the relevant sequences of input variables;
   - each internal state variable is assigned, in general, to an S–R flip-flop;
   - the output variables are generated from the internal state variables or from these and the input variables.

On the other hand, although this method is called 'characterization by levels', it is a mixed method because the reset of the S–R flip-flops can be done by means of a pulse with a duration equal to the PLC working cycle (Figure 4.9).

Figure 5.45a represents, in ladder diagram language, the circuit required to generate a pulse when X1 is opened. This circuit uses two internal state

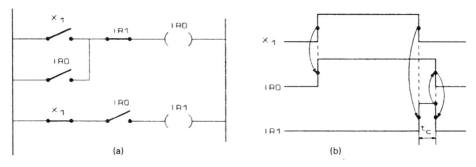

FIGURE 5.45 *Program generating a pulse in internal variable IR1 when X_1 variable switches off.*

variables IR0 and IR1. It is assumed that both IR0 and IR1 are initially open, like X1. When X1 is closed, IR0 closes too but IR1 remains open. When X1 is open, again IR1 is switched on too, and IR0 remains switched on until the next working cycle, at which time it is switched off. The switching-off of IR0 leads to the switching-off of IR1 in the same working cycle. IR1 stays switched on, therefore, only during one working cycle (t_c) of the PLC. Figure 5.45b shows the evolution of X1, IR0 and IR1 in order to illustrate what has just been said.

Figure 5.46 shows a circuit generating a pulse in IR1 when X1 goes from 0 to 1. The reader is advised to analyze its operation.

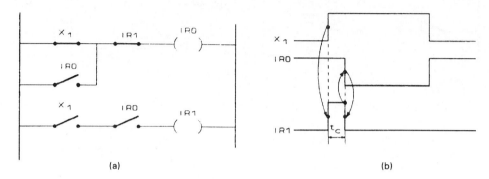

FIGURE 5.46 *Program generating a pulse in internal variable IR1 when X_1 variable switches on.*

The order of the lines in the ladder diagram of Figure 5.45 implies that variable IR0 is generated before IR1 during the PLC processing cycle. If this order is changed, as indicated in Figure 5.47a, a different behaviour results, but the practical outcome is the same (Figure 5.47b).

In a similar way, the lines of Figure 5.46 may be changed, also obtaining a pulse in IR1 when X1 switches on.

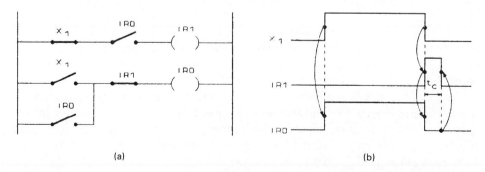

FIGURE 5.47 *Program generating a pulse in internal variable IR1 when X_1 variable switches off.*

The circuits of Figures 5.45, 5.46 and 5.47 may be simplified by eliminating the flip-flop assigned to IR0. Figure 5.48 represents a circuit for the generation of a pulse when X1 switches off, and again the reader is advised to analyze its operation. The circuit for the generation of a pulse when X1 switches on is represented in Figure 5.49.

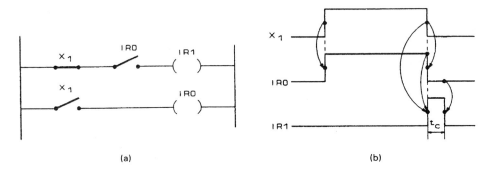

FIGURE 5.48 *Program generating a pulse in internal variable IR1 when X_1 variable switches off.*

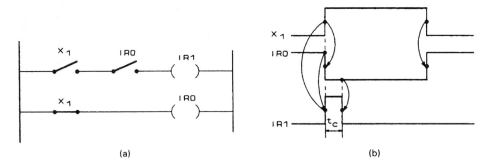

FIGURE 5.49 *Program generating a pulse in internal variable IR1 when X_1 variable switches on.*

A detailed analysis of Figures 5.48 and 5.49 allows us to say that these two contact lines have to be programmed in the order shown because their inversion prevents variable IR1 being switched on because it becomes equivalent to the logic product of X1 and $\overline{X1}$.

On the other hand, it should be emphasized that the order of execution of the instructions in a PLC determines its speed of response to certain input, output or internal state variable transitions. As an example, Figure 5.50 represents two programs in ladder diagram language which execute the same equations but with different response speeds.

In the circuit of Figure 5.50a, when X1, X2 and X3 are closed and X0 changes from open to closed, Y0, Y1, Y2 and Y3 are switched on in the same

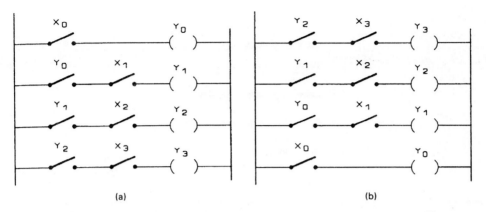

FIGURE 5.50 *Example of programs generating the same equations with a different response time.*

cycle. However, in the circuit of Figure 5.50b, in the same conditions, Y0, Y1, Y2 and Y3 are switched on in successive PLC cycles.

The examples below will contribute to a better understanding of this method. Example 5.3 is based on the design orientated towards output variables, and example 5.4 on the design orientated towards internal state variables.

EXAMPLE 5.3

Design, using a PLC, the automatic control system for the garage of Figure 5.51.

The entrance and exit to the garage are controlled by a pair of barriers driven by electric motors M1 and M2, respectively. On both sides of the barriers two sensors detect the presence of cars: S1 and S2 at the entrance and S3 and S4 at the exit. These sensors remain switched on while there is a car in front of them, and by construction, neither S1 and S2 nor S3 and S4 are switched on simultaneously. A sensor S5 is used to detect the ticket at the exit control.

The garage has capacity for ten cars and the electronic system should control the following actions:

1. Automatic opening and closing of the barriers. The entrance barrier should open if the garage has less than ten cars and a rising edge (transition from 0 to 1) of sensor S1 output is produced. This barrier should close if a falling edge (transition from 1 to 0) of sensor S2 is produced.

    The exit barrier should open if S5 is switched on (a ticket is inserted) and a rising edge in S3 is produced. It should close when a falling edge in S4 is produced.

2. An entrance sign green lamp GL, indicating there are spaces available in the garage.

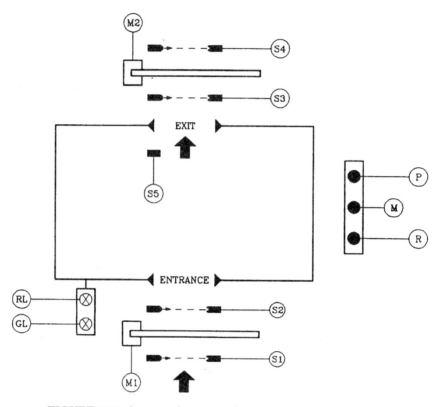

FIGURE 5.51 *Layout of a garage being controlled using a PLC.*

3. An entrance sign red lamp RL, indicating the garage is full.

    The control system should also have the following inputs:
    - a start-up push-button M – from the moment power is supplied to the PLC, no entries or exits are permitted until this push-button is switched on;
    - a stop push-button P – if this push-button is switched on, the entrance and exit of cars becomes impossible until push-button M is again switched on. In the case where P and M are simultaneously switched on, P prevails over M;
    - a push-button R to reset the car counter when the PLC is powered up.

Solution

To begin with, we must assign input variables X to the different external input variables and output variables Y to the different external output variables. This is done in Table 5.6. Then, the design can be as follows:

1. Since the system has a start M and a stop P push-buttons, implementation must be with an asynchronous S–R flip-flop with priority stop, whose ladder diagram is shown in Figure 5.52. This flip-flop constitutes the internal state variable IR0.

PLC PROGRAMMING LANGUAGES

TABLE 5.6

External variable	PLC of section 5.2 variable assignment
S1	X0
S2	X1
S3	X2
S4	X3
S5	X4
M	X5
P	X6
R	X7
M1	Y0
M2	Y1
GL	Y2
RL	Y3

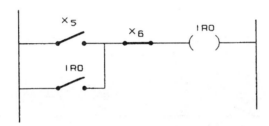

FIGURE 5.52 *Program corresponding to the S–R flip-flop which stores the activation of start and stop push-buttons.*

2. Implementation and assignment of an asynchronous S–R flip-flop to the output variable Y0 which controls motor M1. This variable should be switched on when S1 (X0) is 1 provided IR0 is a 1 and should remain switched on when S1 becomes a 0. We obtain the circuit of Figure 5.53. In addition to what has been said, it is also necessary that M1 switches off when S2 changes from 1 to 0 on the entrance of a car into the garage. For that, it is necessary to generate a pulse by means of two internal state variables. We use IR1 and IR2 according to the theory explained above. The corresponding circuit is shown in Figure 5.54 and the reader may verify that IR2 is switched on during an input and output cycle if X1 is switched off while Y0 is switched on. For the pulse appearing in IR2 to switch off Y0, we need to modify the circuit of Figure 5.53 as in Figure 5.55.

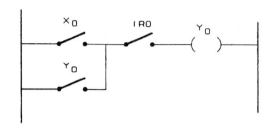

FIGURE 5.53 *Provisional program corresponding to S–R flip-flop assigned to the output variable Y0 which controls motor M1.*

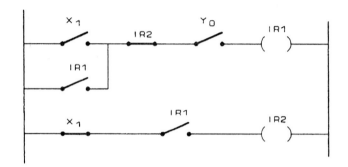

FIGURE 5.54 *Program generating a pulse in variable IR2 when variable S2 (X1) switches off.*

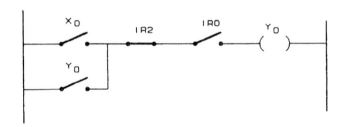

FIGURE 5.55 *Provisional program obtained from Figure 5.53 corresponding to S–R flip-flop assigned to the output variable Y0.*

3. Implementation and assignment of an asynchronous S–R flip-flop (IR3) to remember the insertion of a ticket when S5 (X4) becomes a 1. This flip-flop can only be switched on if IR0 is also switched on, and is switched off when the stop button is switched on. The corresponding circuit is that of Figure 5.56.

4. Implementation and assignment of an S–R flip-flop to the output variable Y1 (M2). This flip-flop should be switched on by X2 (there is a car in front of S3) provided IR3 is also switched on (Figure 5.57).

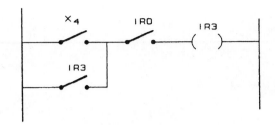

FIGURE 5.56 *Provisional program corresponding to S–R flip-flop assigned to variable IR3 which stores the insertion of a ticket using input variable S5.*

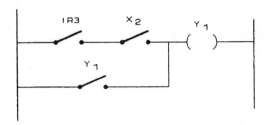

FIGURE 5.57 *Provisional program corresponding to S–R flip-flop assigned to the output variable Y1 which controls motor M2.*

5. Both flip-flops IR3 and Y1 should be switched off when a car comes out of the exit. This action is detected by X3 switching off (S4). For that, two internal state variables are used, IR4 and IR5, which are connected according to the circuit of Figure 5.58. IR5 switches on during a cycle if X3 switches on, provided Y1 is active. Also, IR0 should be active. The circuits of Figures 5.56 and 5.57 become that of Figure 5.59.

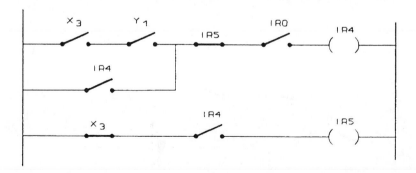

FIGURE 5.58 *Program generating a pulse in variable IR5 when variable S4 (X3) switches on.*

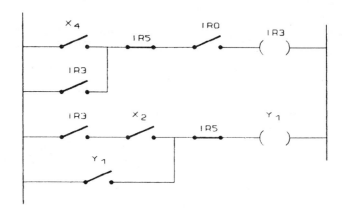

FIGURE 5.59 *Final program corresponding to S–R flip-flops assigned to variables IR3 and Y1.*

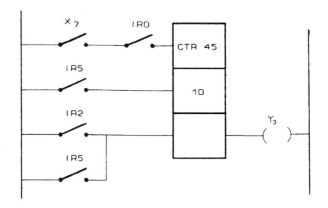

FIGURE 5.60 *Program corresponding to the up/down-counter of cars.*

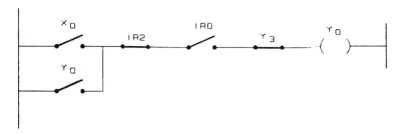

FIGURE 5.61 *Final program corresponding to S–R flip-flop assigned to variable IR0.*

PLC PROGRAMMING LANGUAGES

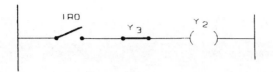

FIGURE 5.62 *Program corresponding to output variable Y2 which controls the green lamp.*

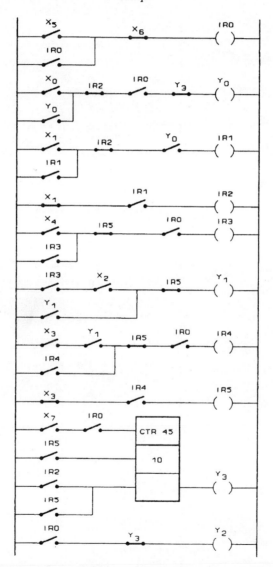

FIGURE 5.63 *Program to control the garage of Figure 5.51 using the ladder diagram language of the PLC described in section 5.2.*

TABLE 5.7

Instruction list		Comment
STR X5 OR IR0 AND NOT X6 OUT IR0	}	Start-stop S-R flip-flop
STR X0 OR Y0 AND NOT IR2 AND IR0 AND NOT Y3 OUT Y0	}	S-R flip-flop assigned to Y0(M1)
STR X1 OR IR1 AND NOT IR2 AND Y0 OUT IR1 STR NOT X1 AND IR1 OUT IR2	}	Generation of a pulse in IR2 when X1 switches on
STR X4 OR IR3 AND NOT IR5 AND IR0 OUT IR3	}	S-R flip-flop assigned to ticket store
STR IR3 AND X2 OR Y1 AND NOT IR5 OUT Y1	}	S-R flip-flop assigned to Y1(M2)
STR X3 AND Y1 OR IR4 AND NOT IR5 AND IR0 OUT IR4 STR NOT X3 AND IR4 OUT IR5	}	Generation of a pulse in IR5 when X3 switches on
STR X7 AND IR0 STR IR5 STR IR2 OR IR5 CTR 45 10 OUT Y3	}	Cars counting, detection of content equals 10, and RL generation
STR IR0 AND NOT Y3 OUT Y2 END	}	GL generation

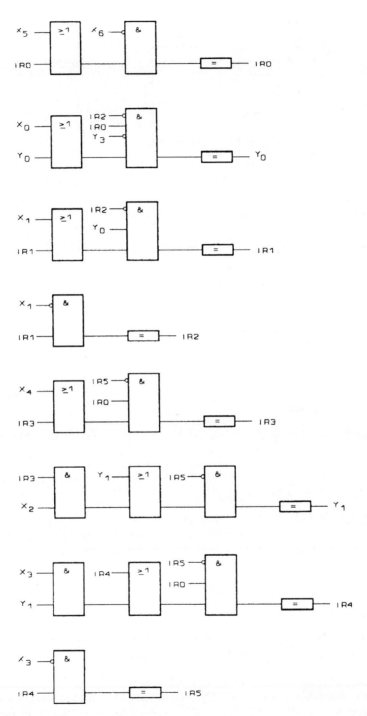

FIGURE 5.64 *Program to control the garage of Figure 5.51 using the function diagram language of the PLC described in section 5.2 (continued overleaf).*

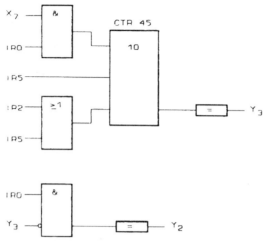

FIGURE 5.64 *Continued.*

6. The car counter activates Y3 whenever the number of cars inside the garage is ten. This counter should be up/down and act according to what was said in section 5.3.2 (Figure 5.21). Its ladder diagram is represented in Figure 5.60. The counter is reset if IR0 is active and R (X7) becomes active. It counts up IR2 pulses if IR5 is switched off and down in the opposite case, and receives as counting pulses IR2 (up) and IR5 (down).

7. Y3 controls the red lamp (RL) and the activation of Y0 (M1). Consequently the circuit of Figure 5.55 should be converted into that of Figure 5.61.

8. The green lamp GL should light up if IR0 is switched on and Y3 is not. Its circuit is shown in Figure 5.62.

Bringing together all the circuits just designed we arrive at the PLC program in ladder diagram language shown in Figure 5.63. From that the reader can immediately obtain the program in instruction list and function diagram languages represented in Table 5.7 and Figure 5.64, respectively.

EXAMPLE 5.4

Design with a PLC the automatic control system of the car-washing machine of Figure 5.65. The system has the following elements:

1. Three motors executing the following tasks:
    - the main motor (MP) which moves the machine along the track and has two control variables MP1 and MP2. When MP1 is switched on the machine

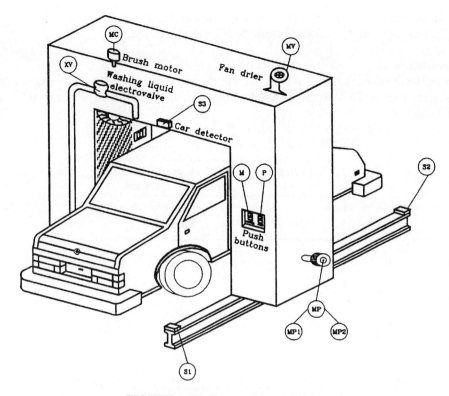

FIGURE 5.65 *Car-washing machine.*

moves from right to left, and it moves in the opposite direction when MP2 is switched on;
- motor of the brushes (MC);
- motor of the fan (MV).
2. An electrovalve (XV) which controls washing liquid on the car.
3. A sensor S3 detecting the presence of a car.
4. Two limit switches S1 and S2 detecting the arrival of the machine at the end of the track.

The machine should operate as follows:

1. Initially the machine is on the right end of the track (S2 switched on) and should start when push-button M is switched on and a car is inside it (S3 switched on).
2. Once M is switched on the machine should do a round trip with the washing liquid output open and the brushes working.
3. When the machine reaches the right end (S2 is again switched on) another round trip should take place but now only the fan should be working. Once this trip is over the machine should stop and return to the initial position.

4. In an emergency, the stop push-button P should be switched on so that the cycle is interrupted and the machine returns to the initial position.

Solution

First, the input variables X are assigned to the different external input variables and the output variables Y to the different output variables. This assignment is indicated in Table 5.8.

TABLE 5.8

External variable	PLC of section 5.2 variable assignment
S1	X0
S2	X1
S3	X2
M	X3
P	X4
MP1	Y0
MP2	Y1
MV	Y2
MC	Y3
XV	Y4

Next, the design of the control program is done in ladder diagram language by the trial-and-error method described in section 5.6.3, using the variant orientated towards the internal state variables. The result obtained is shown in Figure 5.66 and is described next.

Variable IR0 stores the action on the start button M (X3). It consists of a priority reset S–R flip-flop. This flip-flop is switched on when the start button M (X3) is depressed if at that time the machine is at the right end, which is indicated by limit switch S2 (X1) being switched on. The reset of IR0 takes place in any of the following circumstances:

1. The stop button P (X4) is depressed.
2. There is no car in the machine. This situation is indicated by S3 (X2) being switched off.
3. Variable IR4 is switched on which indicates that the second trip from left to right is over (to be analyzed later).

Variable IR1 is an S–R flip-flop switched on when limit switch S1 (X0) turns on (when the machine reaches the left end). Therefore, IR1 indicates that the machine has finished the first trip from right to left. IR1 is reset when IR0 is switched off. The inverse of variable S2 (X1) is placed in series with S1 (X0)

PLC PROGRAMMING LANGUAGES   199

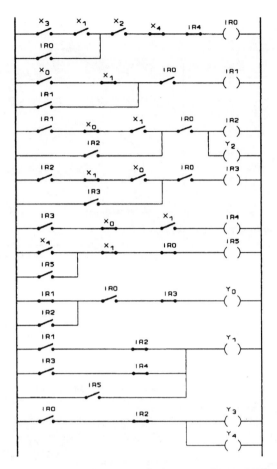

FIGURE 5.66 *Program to control the car-washing machine of Figure 5.65 using the instruction list language of the PLC described in section 5.2.*

as a safety measure to prevent the activation of IR1 if the limit switch X1 is closed.

Variable IR2 is an S–R flip-flop switched on when limit switch S2(X1) is closed if IR1 is active (the first trip from right to left has already been done). Therefore, its activation indicates the first trip from the left to the right is over. IR2 is also reset when IR0 is switched off. The inverse of X1 is placed in series with X0 also as a safety measure.

Variable IR3 is another S–R flip-flop switched on when limit switch S1 (X0) is closed if IR2 is active (which indicates that the first trip from left to right has already been done). Therefore, its activation indicates that the second trip from right to left is done. IR3 is reset when IR0 is switched off. Again, the inverse of X1 is placed in series with X0 for safety reasons.

TABLE 5.9

Instruction list		Comment
STR	X3	
AND	X1	
OR	IR0	
AND	X2	Start-stop S-R flip-flop
AND NOT	X4	
AND NOT	IR4	
OUT	IR0	
STR	X0	
AND NOT	X1	
OR	IR1	S-R flip-flop which sets when S1
AND	IR0	limit switch turn on
OUT	IR1	
STR	IR1	
AND NOT	X0	
AND	X1	
OR	IR2	S-R flip-flop which sets when S2
AND	IR0	limit switch turn on being IR1=1
OUT	IR2	
OUT	Y2	
STR	IR2	
AND NOT	X1	
AND	X0	
OR	IR3	S-R flip-flop which sets when S1
AND	IR0	limit switch turn on being IR2=1
OUT	IR3	
STR	IR3	
AND NOT	X0	S-R flip-flop which sets when S2
AND	X1	limit switch turn on being IR3=1
OUT	IR4	
STR	X4	
OR	IR5	S-R flip-flop which stores the
AND NOT	X1	stop command (P)
AND NOT	IR0	
OUT	IR5	
STR NOT	IR1	
OR	IR2	
AND	IR0	Generation of output Y0(M1)
AND NOT	IR3	
OUT	Y0	
STR	IR1	
AND NOT	IR2	
STR	IR3	
AND NOT	IR4	
OR STR		Generation of output Y1(M2)
STR	IR5	
OR STR		
OUT	Y1	
STR	IR0	
AND NOT	IR2	Generation of output BM(Y3) and
OUT	Y3	XV(Y4)
OUT	Y4	

PLC PROGRAMMING LANGUAGES

Variable IR4 is switched on when limit switch S2 is closed if IR3 is active (which indicates that the second trip from right to left has already been done). The activation of IR4 resets IR0 which in turn resets IR1, IR2 and IR3.

In order to drive the machine automatically to the right end when P is switched on it is necessary to assign an internal state variable IR5 to store such action. IR5 is an S–R flip-flop switched on when P (X4) is depressed and switched off when X1 (S2) is switched on or when IR0 is also switched on.

Next, the output variables are generated from the input and the internal state variables. For this, it is necessary to analyze the problem statement carefully. In this particular case they depend only on the latter.

The output MP1 (variable Y0) will switch on if the inverse of IR1 or IR2 is activated at the same time as IR0 is activated (start cycle initiated) but IR3 is not activated.

Output MP2 (variable Y1) should be switched on in any of the following situations:

1. If IR1 is switched on and IR2 is not.
2. If IR3 is switched on and IR4 is not.
3. If IR5 is switched on.

Output MV (variable Y2) should be switched on only during the second return trip. Therefore it is identical to the internal state variable IR2.

Output MB (variable Y3) and output XV (variable Y4) should be switched on only during the first return trip. Therefore, they are identical and should be switched on if IR0 is active and IR2 is not.

From Figure 5.66 the reader may easily obtain the program using the instruction list and the function diagram languages shown in Table 5.9 and Figure 5.67, respectively.

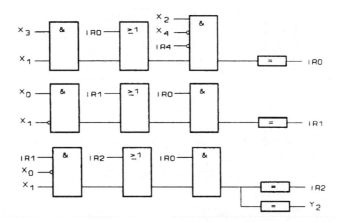

FIGURE 5.67 *Program to control the car-washing machine of Figure 5.65 using the function diagram language of the PLC described in section 5.2 (continued overleaf).*

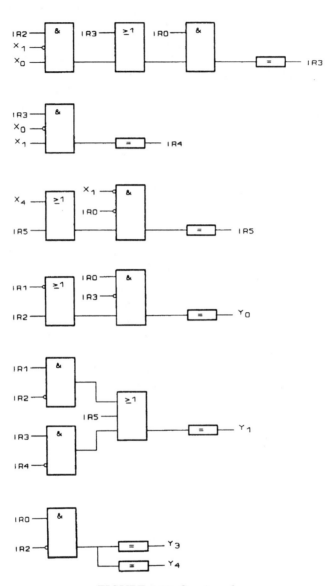

FIGURE 5.67 *Continued.*

Bibliography

[GIRA 73] P. Girard and P. Naslin, *Construction des machines sequentielles industrielles*. Dunod, 1973.

[MAND 91] E. Mandado, *Sistemas electrónicos digitales*, 7th edition, section 6.2.3. Marcombo, 1991.

CHAPTER 6

Commercially available PLCs

6.1 Introduction

The large application field of PLCs has made the design and manufacture of PLCs an attractive proposition to a large number of electronic equipment manufacturers. Furthermore, the many ways in which a PLC may be implemented (chapters 4 and 7) have meant not only that PLCs from different manufacturers have different hardware and software but also that each manufacturer markets several models with different input/output capacity, modularity, communication resources, etc. In this chapter we study the Simatic S5-100U from Siemens and Sysmac C-28K from Omrom.

Throughout this chapter we describe the most important general characteristics of these PLCs and their logic processing instructions. These two PLCs are implemented using microprocessors and therefore they also have numerical information processing capacity, analog variables, etc., which are studied in chapter 7.

6.2 Simatic S5-100U

Simatic S5-100U is a PLC from Siemens with a totally modular configuration. It consists of:

1. One power supply.
2. One central unit that includes reprogrammable read-only memory modules (RPROM) using EPROM (electrically programmable read-only memories) or EEPROM (electrically erasable programmable read-only memories).
3. I/O modules.
4. I/O simulation modules.
5. Special modules.

COMMERCIALLY AVAILABLE PLCS 205

FIGURE 6.1 *Simatic S5-100U PLC (courtesy of Siemens)*.

All the modules are connected to a bus and all of them, together with the central unit and the power supply, are fixed to a rail (Figure 6.1).

6.2.1 General characteristics

The principal characteristics of the Simatic S5-100U PLC are the following:

- Memory capacity: 1024 instructions
- Passive memory: EPROM and EEPROM
- Execution time of each binary operation: 70 µs
- Cycle watching time: 300 ms

- Internal state variables: 1024, of which 512 are non-volatile, i.e. they keep the data when there is a power failure
- Timers: 16
- Timer margin: 0.01 to 9990 s
- Counters: 16, of which 8 are non-volatile
- Counting margin: 0–999 (ascending/descending)
- Digital I/O: 128 as a maximum
- Battery: lithium (3.4 V/850 mA-h)
- Battery lifetime: 5 years
- Central unit power supply: 24 V/0.8 A
- Configuration of the digital input modules:
    4/8 inputs 24 V DC/7 mA
    4 inputs 24–60 V DC/7.5 mA
    4/8 inputs 115 V AC/10 mA
    4/8 inputs 230 V AC/15 mA
- Configuration of the digital output modules:
    4 outputs 24 V DC/0.5 A
    4 outputs 24 V DC/2 A
    8 outputs 24 V DC/0.5 A
    4 outputs 24–60 V DC/0.5 A
    4 outputs 115–230 V AC/1 A
    8 outputs 150–230 V AC/0.5 A
    4 relay outputs 30 V DC/230 V AC

6.2.2 Programming

The S5-100U is implemented using a microprocessor and has two programming modes: linear and structured.

Linear programming is carried out using a programming module called PB1. The instruction sequence of this module is continuously processed by the PLC. In this programming mode only one program module (PB1) is possible, just as in the different basic PLCs of chapter 4 or in the generic PLC of chapter 5.

Structured programming entails dividing the program into several modules and is described in chapter 7.

Variable identification

Each variable (input, internal state or output) is identified by a letter and two digits separated by a decimal point. The letter assigned to each type of variable is one of:

    I: Input variable
    Q: Output variable
    F: Internal state variable

COMMERCIALLY AVAILABLE PLCS 207

In the cases of input and output variables the first digit indicates the position of the module and the second indicates the variable position inside the module. For example:

I 0.3   Input 3 of module 0
I 5.7   Input 7 of module 5

Besides the above three variable types there are two others corresponding to timers and counters.

Tn: Timer and its associated number
Cn: Counter and its associated number

These two variables are included because the timers and counters of the Simatic S5-100U PLC may behave directly as variables, with no need to act through an external or internal output variable as occurs in the generic PLC described in chapter 5.

Instruction set

In this section the main binary logic instructions of the S5-100U are described.

LOGIC AND OUTPUT INSTRUCTIONS

Logic AND function
This function is indicated by the letter A and is used to specify the first variable of a sequence. Figure 6.2 shows an example of the logic AND operation between variables I 0.0 and F 3.5 using instruction list and ladder diagram languages.

```
    :A   I    0.0              !I 0.0      F 3.5
    :A   F    3.5              +---] [---+---] [---+---------+
                               !
         (a)                          (b)
```

FIGURE 6.2 *Logic AND function example.* (a) *Instruction list language.* (b) *Ladder diagram language.*

Logic OR function
This function is indicated by the letter O. It is always used with the logic AND (A). Figure 6.3a shows, as an example, the logic OR operation between variables I 2.3, I 0.1, F 3.2 and Q 1.1 using the instruction list language. Figure 6.3b shows its equivalent using the ladder diagram language.

Inversion logic function
The inversion of a variable is indicated by letter N. If an inverted variable is the first of a sequence or is part of an AND function, code AN is used, and if

```
                              !
                              !I 2.3
        :A    I    2.3        +---] [---+---------+
                              !         !
                              !I 0.1    !
        :O    I    0.1        +---] [---+
                              !         !
                              !F 3.2    !
        :O    F    3.2        +---] [---+
                              !         !
                              !Q 1.1    !
        :O    Q    1.1        +---] [---+
                              !
              (a)                    (b)
```

FIGURE 6.3 *Logic OR function example.* (a) *Instruction list language.* (b) *Ladder diagram language.*

```
                              !
                              !I 0.0
       :AN    I    0.0        +---]/[---+---------+
                              !         !
                              !I 0.1    !
       :ON    I    0.1        +---]/[---+
                              !
              (a)                    (b)
```

FIGURE 6.4 *Inversion logic function example.*

it is part of an $\overline{\text{OR}}$ function code ON is used. Figure 6.4 represents the expression $\overline{I\ 0.0} + \overline{I\ 0.1}$.

Output function (internal or external)
An output variable, being external or internal, is indicated by symbol =. Figure 6.5 represents the transfer of the state of input variable I 0.0 to the output Q 1.0.

By the combination of the four functions just studied, any logic AND operation, logic OR operation or combination of both and the assignment of the result obtained to an output variable may be executed. Figure 6.6 shows the program corresponding to the equation:

$$Q\ 1.6 = I\ 3.0 \cdot \overline{I\ 0.5} \cdot \overline{F\ 4.7} \cdot Q\ 1.0$$

```
       :A    I    0.0        !
       :=    Q    1.0        !I 0.0                    Q 1.0
                             +---] [---+-----------+--(   )-!
                             !
              (a)                    (b)
```

FIGURE 6.5 *Transfer of the state of an input variable to an output variable.*

```
:A    I    3.0                !
:AN   I    0.5                !I 3.0      I 0.5      F 4.7      Q 1.0      Q 1.6
:AN   F    4.7                +---] [---+---]/[---+---]/[---+---] [---+--(    )-!
:A    Q    1.0                !
:=    Q    1.6                !


     (a)                                              (b)
```

FIGURE 6.6 *Assignment of the state of a logic AND function to an output variable.*

```
                                   !
                                   !I 2.3                    Q 5.7
     :A    I    2.3                +---] [---+----------+--(    )-!
                                   !         !
                                   !I 2.5    !
     :O    I    2.5                +---] [---+
                                   !         !
                                   !F 5.0    !
     :ON   F    5.0                +---]/[---+
                                   !         !
                                   !I 3.2    !
     :ON   I    3.2                +---]/[---+
                                   !         !
                                   !Q 5.2    !
     :O    Q    5.2                +---] [---+
                                   !
     :=    Q    5.7


              (a)                              (b)
```

FIGURE 6.7 *Assignment of the state of a logic OR function to an output variable.*

which is equivalent to the combination of serial contacts. Figure 6.7 shows the program corresponding to the equation:

$$Q\,5.7 = I\,2.3 + I\,2.5 + \overline{F\,5.0} + \overline{I\,3.2} + Q\,5.2$$

which is equivalent to the combination of parallel contacts. Figure 6.8 represents the parallel combination of serial contacts which is equivalent to a logic OR function of logic AND functions and corresponds to the equation:

$$\begin{aligned}Q\,6.4 = &\ I\,0.5 \cdot \overline{I\,0.2} \cdot F\,4.3 \cdot Q\,6.0 \cdot \overline{T\,4} \cdot F\,2.6 \\ &+ I\,0.1 \cdot \overline{F\,3.2} \cdot I\,0.4 \cdot \overline{T\,1} \cdot \overline{C\,9} \cdot \overline{Q\,6.1} \\ &+ I\,1.1 \cdot \overline{F\,2.7} \cdot \overline{C\,0} \cdot \overline{Q\,9.1} \cdot Q\,8.7 \cdot T\,9\end{aligned}$$

Figure 6.9 represents the serial combination of parallel contacts which is equivalent to a logic AND function of logic OR functions and corresponds to the equation:

$$Q\,5.5 = (I\,0.0 + I\,0.1)(I\,2.3 + Q\,5.2)(F\,1.6 + I\,2.7)$$

Each parallel set is separated from the others by parentheses which begin and end its specification.

More complex combinations may be performed, like the one represented in Figure 6.10 corresponding to the equation:

$$Q\,2.0 = (I\,0.0 \cdot \overline{F\,0.0} + \overline{I\,0.1} \cdot C\,2)(\overline{I\,0.7} \cdot T\,6 \cdot \overline{Q\,1.0} + F\,3.2 \cdot I\,0.3)I\,0.5$$

```
!I 0.5      I 0.2      F 4.3      Q 6.0      T 4        F 2.6                           Q 6.4
+---] [---+---]/[---+---] [---+---] [---+---]/[---+---] [---+----------+--(   )-!
!                                                                 !
!I 0.1      F 3.2      I 0.4      T 1        C 9        Q 6.1     !
+---] [---+---]/[---+---] [---+---]/[---+---]/[---+---]/[---+
!                                                                 !
!I 1.1      F 2.7      C 0        Q 9.1      Q 8.7      T 9       !
+---] [---+---]/[---+---]/[---+---]/[---+---] [---+---] [---+
!
```

(a)

```
002E    :A     I    0.5
0030    :AN    I    0.2
0032    :A     F    4.3
0034    :A     Q    6.0
0036    :AN    T    4
0038    :A     F    2.6
003A    :O
003C    :A     I    0.1
003E    :AN    F    3.2
0040    :A     I    0.4
0042    :AN    T    1
0044    :AN    C    9
0046    :AN    Q    6.1
0048    :O
004A    :A     I    1.1
004C    :AN    F    2.7
004E    :AN    C    0
0050    :AN    Q    9.1
0052    :A     Q    8.7
0054    :A     T    9
0056    :=     Q    6.4
```

(b)

FIGURE 6.8 *Example of a logic OR function of logic AND functions.*

```
                                                           :A(
                                                           :O    I    0.0
                                                           :O    I    0.1
!                                                          :)
!I 0.0      I 2.3      F 1.6                Q 5.5          :A(
+---] [---+---] [---+---] [---+----------+--(   )-!        :O    I    2.3
!           !          !          !                        :O    Q    5.2
!I 0.1     !Q 5.2     !I 2.7      !                        :)
+---] [---+---] [---+---] [---+                            :A(
!                                                          :O    F    1.6
                                                           :O    I    2.7
                                                           :)
                                                           :=    Q    5.5
```

(a)   (b)

FIGURE 6.9 *Example of a logic AND function of logic OR functions.*

COMMERCIALLY AVAILABLE PLCS

```
!
!I 0.0      F 0.0       I 0.7      T 6        Q 1.0      I 0.5                 Q 2.0
+---] [---+---]/[---+---]/[---+---] [---+---]/[---+---] [---+----------+--(    )-!
!                       !                                !
!I 0.1       C 2       !F 3.2      I 0.3                 !
+---]/[---+---] [---+---] [---+---] [---+----------+
!
                                    (a)
```

```
            000A    :A(
            000C    :A    I    0.0
            000E    :AN   F    0.0
            0010    :O
            0012    :AN   I    0.1
            0014    :A    C    2
            0016    :)
            0018    :A(
            001A    :AN   I    0.7
            001C    :A    T    6
            001E    :AN   Q    1.0
            0020    :O
            0022    :A    F    3.2
            0024    :A    I    0.3
            0026    :)
            0028    :A    I    0.5
            002A    :=    Q    2.0
            002C    :BE
```

(b)

FIGURE 6.10 *Example of a complex logic function which combines logic AND and logic OR functions using direct and inverted variables.*

TIMER FUNCTION

The timers of the S5-100U PLC have a timing input variable, a reset input variable and a timing output variable. The time is programmed by instruction L KTXXX.Y where:

    XXX   represents the relative value (0 to 999)
    Y   represents the time base; it is a factor varying from 0 to 3 and multiplies the relative value by 0.01, 0.1, 1 or 10, respectively, to obtain the time in seconds.

Because the S5-100U has numerical processing capacity, it has operands of different dimensions (see section 7.4.1) as well as coded load instructions (LC) and transfer T instructions which allow the time value to be displayed (see section 7.4.1).

There are five types of timer, studied next. In each case it is assumed that the timing input variable is I 0.0, the reset input variable is I 0.1 and the timing output variable is Q 4.0; to visualize the timing, the word QW 6 is used (see section 7.4.1).

1. **Non-stored pulse triggering (SP)**. In this type of timer, output variable (Q 4.0) is switched on and stays switched on during a maximum time T from the instant a rising edge occurs at the input (change from 0 to 1 of the timing input variable (I 0.0)). If the timing input variable (I 0.0)

returns to 0 before time T has elapsed, the output switches off at the same instant. To begin the timing function the reset input variable (I 0.1) must be at logic 0.

The output variable is switched off at the instant the reset input variable is switched on (logic 1) and is not set again until I 0.1 is at 0 and a new rising edge is applied to I 0.0.

The timing value is externally displayed, executing one coded load instruction (LC) followed by one transfer instruction (T).

Figure 6.11 shows the timing diagram of the three signals. Table 6.1 shows the corresponding program in the instruction list language, and Figure 6.12 shows it in the ladder diagram and the function diagram languages. The 'no operation' instructions (NOP) included in Table 6.1 do not outwardly affect the operation of the PLC: they are included because they are needed by the programs translating from instruction list to ladder and function diagram languages in order to execute the translation.

2. **Stored pulse triggering (SE).** The behaviour of this type of timer is similar to the foregoing except that if the input variable changes to 0 before time T has elapsed, the output stays switched on until the end of that time. This timer is also retriggerable: that is, if while the output is

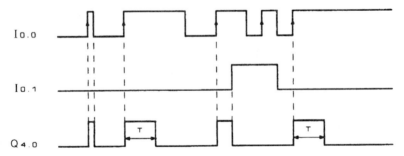

FIGURE 6.11 *Timing diagram of the variables of a non-stored pulse triggering timer (SP).*

TABLE 6.1 *Program of a non-stored pulse triggering timer (SI) using the instruction list language of Simatic S5-100U PLC.*

```
0000    :A     I      0.0
0001    :L     KT   005.2
0003    :SP    T      1
0004    :A     I      0.1
0005    :R     T      1
0006    :NOP   0
0007    :LC    T      1
0008    :T     QW     6
0009    :A     T      1
000A    :=     Q      4.0
000B    :BE
```

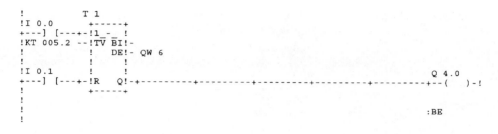

(a)

(b)

FIGURE 6.12 *Program corresponding to a non-stored pulse triggering timer (SP).* (a) *Ladder diagram language.* (b) *Function diagram language.*

switched on several pulses are applied to the timing input variable, then each of them will restart it.

This timer's display is done in the same way as timer SP's.

Figure 6.13 shows the timing diagram of this timer's three signals. Table 6.2 shows the corresponding program in instruction list, and Figure 6.14 shows it in the ladder diagram and function diagram languages.

3. **Switch-on delay (SD)**. The output variable of this timer (Q 4.0) is switched on when time T has elapsed from the instant when the timing input variable (I 0.0) changes from 0 to 1, and stays switched on while the

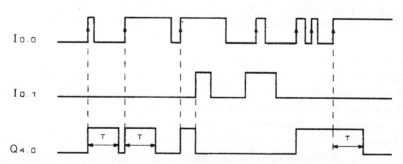

FIGURE 6.13 *Timing diagram of the variables of a stored pulse triggering timer (SE).*

TABLE 6.2 *Program of a stored pulse triggering timer (SV) using the instruction list language of Simatic S5-100U PLC.*

```
0000  :A    I    0.0
0001  :L    KT   050.1
0003  :SE   T    2
0004  :A    I    0.1
0005  :R    T    2
0006  :NOP  0
0007  :LC   T    2
0008  :T    QW   6
0009  :A    T    2
000A  :=    Q    4.0
000B  :BE
```

```
!              T 2
!I 0.0        +-----+
+---] [---+-!1_-_V!
!KT 050.1 --!TV BI!-
!          !   DE!- QW 6
!          !    !                                                                  Q 4.0
!I 0.1     !    !                                                                +--( )-!
+---] [---+-!R   Q!-+---------+----------+----------+----------+----------+--(   )-!
!          +-----+
!
!                                                                                :BE
!
```

(a)

```
                T 2
               +-----+
I 0.0        --!1_-_V!
KT 050.1     --!TV BI!-
               !  DE!- QW 6
               !   !   +------+
I 0.1        --!R  Q!-+-! =   ! Q 4.0
               +----+   +------+
                                :BE
```

(b)

FIGURE 6.14 *Program corresponding to a stored pulse triggering timer (SE). (a) Ladder diagram language. (b) Function diagram language.*

latter is at level 1. If I 0.0 stays switched on for a period less than T, the output variable is not switched on. The output variable stays at level 0 while the reset input variable (I 0.1) is switched on and does not initiate a timing even if a rising edge is applied to I 0.0.

The display for this timer is done in the same way as for timer SP.

Figure 6.15 shows the timing diagram of the three signals of this timer. Table 6.3 shows the corresponding program in the instruction list language, and Figure 6.16 shows it in the ladder diagram and function diagram languages.

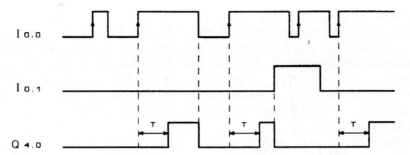

FIGURE 6.15 *Timing diagram of the variables of a switch-on delay timer (SD).*

TABLE 6.3 *Program of a switch-on delay timer (SE) using the instruction list language of Simatic S5-100U PLC.*

```
0000    :A    I     0.0
0001    :L    KT    500.0
0003    :SD   T     3
0004    :A    I     0.1
0005    :R    T     3
0006    :NOP  0
0007    :LC   T     3
0008    :T    QW    6
0009    :A    T     3
000A    :=    Q     4.0
000B    :BE
```

```
!            T 3
!I 0.0       +-----+
+---] [---+-!T!-!0!
!KT 500.0 --!TV BI!-
!         !  DE!- QW 6
!         !    !
!I 0.1    !    !                                                                    Q 4.0
+---] [---+-!R  Q!-+---------+---------+---------+---------+---------+--( )-!
!           +-----+
!
!                                                                                   :BE
!
```

(a)

(b)

FIGURE 6.16 *Program corresponding to a switch-on delay timer (SD). (a) Ladder diagram language. (b) Function diagram language.*

4. **Stored switch-on delay (SS).** The behaviour of this type of timer is similar to the previous one with the difference that the timing continues even if the pulse applied to the timing input variable has a duration lower than the timing value T. This timer is also retriggerable and, therefore, if several consecutive rising edges are applied to I 0.0, the output variable is switched on during a time interval T from the last of them. Once Q 4.0 is switched on, the output variably only changes to 0 if the reset variable (I 0.1) is switched on.

Figure 6.17 shows the timing diagram of the three signals of this timer. Table 6.4 shows the corresponding program in instruction list, and Figure 6.18 shows it in ladder diagram and function diagram.

5. **Switch-off delay (SF).** In this type of timer, the output variable (Q 4.0) is switched on when the timing input variable (I 0.0) switches on, and stays switched on until an interval of T has elapsed from the instant the timing input variable switches off. If several consecutive falling edges are applied to the timing variable (I 0.0), the output variable (Q 4.0) is switched on and stays switched on until an interval of T has elapsed from the last pulse. When the reset variable (I 0.1) switches on, the output variable (Q 4.0) is switched off, but if I 0.1 switches off while I 0.0 is a

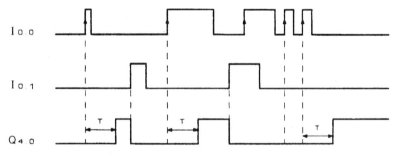

FIGURE 6.17 *Timing diagram of the variables of a stored switch-on delay timer (SS).*

TABLE 6.4 *Program of a stored switch-on delay timer (SS) using the instruction list language of Simatic S5-100U PLC.*

```
0000    :A      I       0.0
0001    :L      KT      005.3
0003    :SS     T       4
0004    :A      I       0.1
0005    :R      T       4
0006    :NOP    0
0007    :LC     T       4
0008    :T      QW      6
0009    :A      T       4
000A    :=      Q       4.0
000B    :BE
```

COMMERCIALLY AVAILABLE PLCS   217

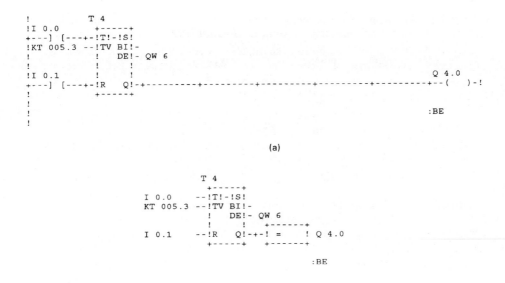

FIGURE 6.18 *Program corresponding to a stored switch-on delay timer (SS).* (a) *Ladder diagram language.* (b) *Function diagram language.*

logic 1, Q 4.0 is immediately switched on and a new switch-off delay is initiated from the instant that I 0.0 returns to state 0.

The display for this timer is done in the same way as for timer SP.

Figure 6.19 shows the timing diagram of the three signals of this timer. Table 6.5 shows the corresponding program in instruction list, and Figure 6.20 shows it using the ladder diagram and function diagram languages.

COUNTER FUNCTION

The counter of Simatic S5-100U is a multifunctional block which can have the following pins:

- One up-counting input variable (CU).

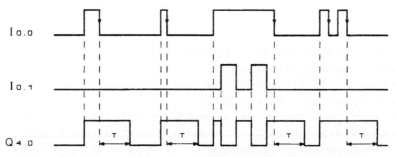

FIGURE 6.19 *Timing diagram of the variables of a switch-off delay timer (SF).*

TABLE 6.5 *Program of a switch-off delay timer (SA) using the instruction list language of Simatic S5-100U PLC.*

```
0000    :A    I    0.0
0001    :L    KT   010.2
0003    :SF   T    5
0004    :A    I    0.1
0005    :R    T    5
0006    :NOP  0
0007    :LC   T    5
0008    :T    QW   6
0009    :A    T    5
000A    :=    Q    4.0
000B    :BE
```

```
!             T 5
!I 0.0        +-----+
+---] [---+-!0!-!T!
!KT 010.2 --!TV BI!-
!         !  DE!- QW 6
!         !   !
!I 0.1    !   !                                                              Q 4.0
+---] [---+-!R  Q!-+----------+----------+----------+----------+----------+--(   )-!
!         +-----+
!
!                                                                                   :BE
!
```

(a)

```
              T 5
              +-----+
I 0.0      --!0!-!T!
KT 010.2   --!TV BI!-
              !  DE!- QW 6
              !   !  +------+
I 0.1      --!R  Q!-+-! =   ! Q 4.0
              +-----+ +------+
                                :BE
```

(b)

FIGURE 6.20 *Program corresponding to a switch-off delay timer (SF). (a) Ladder diagram language. (b) Function diagram language.*

- One down-counting input variable (CD).
- One initial value load variable (S).
- One initial load binary combination (CV).
- One reset input variable (R).
- One display combination in natural binary of the content (BI).
- One display combination in natural BCD of the content (DE).
- One binary output variable which is switched on when the content of the counter is not 0 (Q).

TABLE 6.6 *Program of counter using the instruction list language of Simatic S5-100U PLC (continue).*

```
Segment  1           0000
    :A    I     0.0
    :CU   C     1
    :A    I     0.1
    :CD   C     1
    :A    I     0.2
    :L    KC  005
    :S    C     1
    :A    I     0.3
    :R    C     1
    :L    C     1
    :T    QW    4
    :LC   C     1
    :T    QW    6
    :A    C     1
    :=    Q     4.0
    :***


Segment  2           0011
    :L    C     1
    :L    KF   +6
    :!=F
    :=    Q     4.1
    :***


Segment  3           0017
    :L    C     1
    :L    KF   +7
    :><F
    :=    Q     4.2
    :***


Segment  4           001D
    :L    C     1
    :L    KF   +8
    :>=F
    :=    Q     4.3
    :***


Segment  5           0023
    :L    C     1
    :L    KF   +9
    :>F
    :=    Q     4.4
    :***


Segment  6           0029
    :L    C     1
    :L    KF  +10
    :<=F
    :=    Q     4.5
    :***


Segment  7           002F
    :L    C     1
    :L    KF  +11
    :<F
    :=    Q     4.6
    :***


Segment  8           0035
    :BE
```

```
+------------------------------------------------------------+
! S I E M E N S   S A  -  VIGO (SPAIN) !    UNIVERSITY OF VIGO                      !
!                                      ! DEPARTMENT OF ELECTRONIC TECHNOLOGY        !
+------------------------------------------------------------+
!      PROGRAMMABLE LOGIC CONTROLLER   !                                            !
!           SIMATIC S5 - 100U          !                                    PAG.!
!               SIEMENS PLC            !    DATE: DECEMBER 1994              1!
+------------------------------------------------------------+
```

Because it is a PLC with arithmetic capacity, Simatic S5-100U uses comparison instructions to detect the content of a counter. These instructions are described in section 7.4.1.

Table 6.6 shows the program in the instruction list language corresponding to a counter where all the input and output variables described before are used and where all the possible comparisons are executed. The program is divided into segments corresponding to contact lines in the ladder language. The counter is labelled C1 and the input and output variables are labelled as indicated in Table 6.7.

TABLE 6.7

Variable	SIMATIC S5-100U variable assignment
Up counting input	I 0.0
Down counting input	I 0.1
Initial value load	I 0.2
Reset input	I 0.3
Output set if $(Z1) \neq 0$	Q 4.0
Output set if $(Z1) = 6$	Q 4.1
Output set if $(Z1) > 7$	Q 4.2
Output set if $(Z1) \geq 8$	Q 4.3
Output set if $(Z1) > 9$	Q 4.4
Output set if $(Z1) \leq 10$	Q 4.5
Output set if $(Z1) < 11$	Q 4.6

In the first segment of the program the input variables are assigned and the counter content displayed. The combination which is introduced in parallel by instruction L KCXXX (in this example XXX = 005) is also executed in this first segment. In segments 2 to 7 the different comparisons are done. In a real case only the variables and the necessary comparisons are used. An example is the automatic control of a garage analyzed in section 6.2.3.

The counters can also be represented in ladder and function diagram languages. Figures 6.21 and 6.22 represent in both languages the counter equivalent of that of the instruction list of Table 6.6.

SELF-RETENTION FUNCTION

This function acts like an S–R flip-flop. It may be applied to external or internal output variables and uses instructions S and R. Figure 6.23 shows an example in the instruction list and ladder diagram languages, and refers to a self-retention function of output Q 1.0 with priority reset. By

```
Segment  1         0000
!
!         C 1
!I 0.0             +-----+
+---] [---+-!CU   !
!         !       !
!I 0.1    !       !
+---] [---+-!CD   !
!         !       !
!I 0.2    !       !
+---] [---+-!S    !
!KC 005   --!CV BI!- QW 4
!         !   DE!- QW 6
!         !       !
!I 0.3    !       !                                                           Q 4.0
+---] [---+-!R   Q!-+---------+---------+-----------+-----------+--(   )-!
!         +-----+
!
Segment  2         0011
!
!         +-----+
!C 1      --!!=  F!
!         !       !                                                           Q 4.1
!KF +6    --!   Q!-+---------+---------+-----------+-----------+--(   )-!
!         +-----+
!
Segment  3         0017
!
!         +-----+
!C 1      --!><  F!
!         !       !                                                           Q 4.2
!KF +7    --!   Q!-+---------+---------+-----------+-----------+--(   )-!
!         +-----+
!
Segment  4         001D
!
!         +-----+
!C 1      --!>=  F!
!         !       !                                                           Q 4.3
!KF +8    --!   Q!-+---------+---------+-----------+-----------+--(   )-!
!         +-----+
!
Segment  5         0023
!
!         +-----+
!C 1      --!>   F!
!         !       !                                                           Q 4.4
!KF +9    --!   Q!-+---------+---------+-----------+-----------+--(   )-!
!         +-----+
!
Segment  6         0029
!
!         +-----+
!C 1      --!<=  F!
!         !       !                                                           Q 4.5
!KF +10   --!   Q!-+---------+---------+-----------+-----------+--(   )-!
!         +-----+
!
Segment  7         002F
!
!         +-----+
!C 1      --!<   F!
!         !       !                                                           Q 4.6
!KF +11   --!   Q!-+---------+---------+-----------+-----------+--(   )-!
!         +-----+
!
Segment  8         0035
!                                                                              :BE


+-----------------------------------------------------------------------------+
! S I E M E N S   S A  -  VIGO (SPAIN) !         UNIVERSITY OF VIGO           !
!                                      ! DEPARTMENT OF ELECTRONIC TECHNOLOGY  !
+-----------------------------------------------------------------------------+
!     PROGRAMMABLE LOGIC CONTROLLER    !                                      !
!          SIMATIC S5 - 100U           !                               PAG.!
!             SIEMENS PLC              !   DATE: DECEMBER 1994            1!
+-----------------------------------------------------------------------------+
```

FIGURE 6.21 *Program of a counter using the ladder diagram language of Simatic S5-100U PLC.*

```
Segment  1         0000
                 C 1
                +-----+
   I 0.0      --!CU  !
   I 0.1      --!CD  !
   I 0.2      --!S   !
   KC 005     --!CV BI!- QW 4
                !  DE!- QW 6
                !   !   +------+
   I 0.3      --!R  Q!-+-! =   ! Q 4.0
                +-----+   +------+

Segment  2         0011
                +-----+
   C 1        --!!=  F!
                !   !   +------+
   KF +6      --!   Q!-+-! =   ! Q 4.1
                +-----+   +------+

Segment  3         0017
                +-----+
   C 1        --!><  F!
                !   !   +------+
   KF +7      --!   Q!-+-! =   ! Q 4.2
                +-----+   +------+

Segment  4         001D
                +-----+
   C 1        --!>=  F!
                !   !   +------+
   KF +8      --!   Q!-+-! =   ! Q 4.3
                +-----+   +------+

Segment  5         0023
                +-----+
   C 1        --!>   F!
                !   !   +------+
   KF +9      --!   Q!-+-! =   ! Q 4.4
                +-----+   +------+

Segment  6         0029
                +-----+
   C 1        --!<=  F!
                !   !   +------+
   KF +10     --!   Q!-+-! =   ! Q 4.5
                +-----+   +------+

Segment  7         002F
                +-----+
   C 1        --!<   F!
                !   !   +------+
   KF +11     --!   Q!-+-! =   ! Q 4.6
                +-----+   +------+

Segment  8         0035
                              :BE

+-------------------------------------------------------------------+
! S I E M E N S   S A  -  VIGO (SPAIN) !       UNIVERSITY OF VIGO   !
!                                      ! DEPARTMENT OF ELECTRONIC TECHNOLOGY !
+-------------------------------------------------------------------+
!        PROGRAMMABLE LOGIC CONTROLLER    !                         !
!             SIMATIC S5 - 100U           !                    PAG.:!
!                SIEMENS PLC              !  DATE: DECEMBER 1994  1!
+-------------------------------------------------------------------+
```

FIGURE 6.22 *Program of a counter using function diagram language of Simatic S5-100U PLC.*

COMMERCIALLY AVAILABLE PLCS

```
:A   I   0.0
:S   Q   1.0
:A   I   0.1
:R   Q   1.0
:A   Q   1.0
:=   Q   1.0
:BE
```

(a)

```
!              Q 1.0
!I 0.0       +-----+
+---] [---+-!S      !
!         !         !
!I 0.1    !         !         Q 1.0
+---] [---+-!R  Q!-+----------+--( )-!
!            +-----+
!
!                                 :BE
!
```

(b)

FIGURE 6.23 *Priority reset S–R flip-flop or self-retention function.*

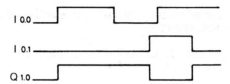

FIGURE 6.24 *Timing diagram of the variables of a priority reset S–R flip-flop.*

programming instruction S Q 1.0 after R Q 1.0 the instruction becomes a priority set self-retention. Figure 6.24 shows the timing diagram of the corresponding signal.

CONTROL INSTRUCTIONS

Control instructions provide decision-making capabilities about whether to execute certain instructions. Simatic S5-100U uses the same kinds of control instruction as computers. Hence, it has conditional (JC) and unconditional (JU) jump instructions which specify the address of the instruction to which the PLC should jump without executing the intermediate instructions.

In the instruction list language of this PLC, the address of the jump instructions constitutes a label identified by character M and a number between 0 and 99.

The conditional jump is combined with the logic and output instructions described above. A simple example is the following instruction sequence:

```
HEX
000  A I 0.0
002  =Q 2.0
004  JC=M0
006  A Q 2.0
008  =Q 2.1
00A  A I 1.0
00C  =Q 2.2
     M0  Continues the sequence
```

In this example, instructions A I 0.0 and =Q 2.0 transfer the state of the input variable I 0.0 to the output Q 2.0. Then, conditional jump instruction JC is executed and, if a 0 has been transferred to Q 2.0, instructions between addresses 006 to 00C are executed; otherwise the PLC does not execute them and jumps to the address indicated by the label M0.

The labels cannot be separated from the jump instruction by more than 127 instructions. When they are separated by more than 127 instructions then intermediate labels must be used. The labels can only be programmed using functional blocks (described in chapter 7).

6.2.3 Practical examples of digital system implementation using the Simatic S5-100U

Some of the control systems described in chapters 1 and 5 are designed next, using the Simatic S5-100U PLC.

Combinational system implementation using the Simatic S5-100U

EXAMPLE 6.1

Implement using Simatic S5-100U PLC the supervising system of the process described in example 3.1.

TABLE 6.8

External variable	SIMATIC S5-100U variable assignment
LSL	I 0.0
TSL	I 0.1
DPSH	I 0.2
PSL	I 0.3
PSH	I 0.4
BS1	I 0.5
BS2	I 0.6
PP	Q 1.0
XV1	Q 1.1
XV2	Q 1.2
XV3	Q 1.3
GL	Q 1.4
RL	Q 1.5

COMMERCIALLY AVAILABLE PLCS

TABLE 6.9

```
Instruction list              Comment
Segment 1     0000

0000 : A    I 0.0  ⎫
0001 : O    I 0.1  ⎪
0002 : O    I 0.2  ⎪
0003 : O    I 0.3  ⎬  PP variable generation
0004 : O           ⎪
0005 : A    I 0.5  ⎪
0006 : A    I 0.6  ⎪
0007 : =    Q 1.0  ⎪
0008 : ***         ⎭

Segment 2     0009

0009 : AN   I 0.5  ⎫
000A : =    Q 1.1  ⎬  XV1 variable generation
000B : ***         ⎭

Segment 3     000C

000C : AN   I 0.6  ⎫
000D : =    Q 1.2  ⎬  XV2 variable generation
000E : ***         ⎭

Segment 4     000F

000F : A    Q 1.0  ⎫
0010 : O    I 0.4  ⎪
0011 : =    Q 1.3  ⎬  XV3 variable generation
0012 : ***         ⎭

Segment 5     0013

0013 : AN   Q 1.0  ⎫
0014 : =    Q 1.4  ⎬  GL variable generation
0015 : ***         ⎭

Segment 6     0016

0016 : A    Q 1.0  ⎫
0017 : =    Q 1.5  ⎬  RL variable generation
0018 : BE          ⎭
```

Solution

The implementation equations given in example 5.1 are the following:

$$PP = LSL + TSL + DPSH + PSL + BS1 \cdot BS2$$
$$XV1 = \overline{BS1}$$
$$XV2 = \overline{BS2}$$

```
Segment 1
!
!I 0.0                                                                                   Q 1.0
+---] [---+---------+-----------+-----------+-----------+-----------+---------+--(   )-!
!         !
!I 0.1    !
+---] [---+---------+
!         !
!I 0.2    !
+---] [---+---------+
!         !
!I 0.3    !
+---] [---+---------+
!         !
!I 0.5      I 0.6   !
+---] [---+---] [---+
!
Segment 2         0009
!
!I 0.5                                                                                   Q 1.1
+---]/[---+---------+-----------+-----------+-----------+-----------+---------+--(   )-!
!
Segment 3         000C
!
!I 0.6                                                                                   Q 1.2
+---]/[---+---------+-----------+-----------+-----------+-----------+---------+--(   )-!
!
Segment 4         000F
!
!Q 1.0                                                                                   Q 1.3
+---] [---+---------+-----------+-----------+-----------+-----------+---------+--(   )-!
!         !
!I 0.4    !
+---] [---+
!
Segment 5         0013
!
!Q 1.0                                                                                   Q 1.4
+---]/[---+---------+-----------+-----------+-----------+-----------+---------+--(   )-!
!
Segment 6         0016
!
!Q 1.0                                                                                   Q 1.5
+---] [---+---------+-----------+-----------+-----------+-----------+---------+--(   )-!
!
!
!                                                                                        :BE
!
```

FIGURE 6.25 *Program of the supervising system of example 6.1 using ladder diagram language of Simatic S5-100U PLC.*

$$XV3 = PP + PSH$$
$$GL = \overline{PP}$$
$$RL = PP$$

We first assign the input and output variables to the PLC variables (Table 6.8). The program in instruction list language is given in Table 6.9 with explanatory comments. Figures 6.25 and 6.26 show the program in ladder diagram and function diagram languages, respectively.

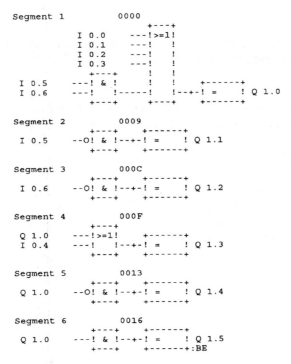

FIGURE 6.26 *Program of the supervising system of example 6.1 using function diagram language of Simatic S5-100U PLC.*

Implementation of edge-characterized sequential control systems using the Simatic S5-100U

To implement asynchronous sequential systems characterized by edges the PLC uses control instructions to decide if internal state changes are to be executed. In this section two control systems implemented with the Simatic S5-100U PLC are analyzed. Conditional jumps to given program addresses specified by labels (as explained in section 6.2.2) are used. As was also explained in section 6.2.2, labels can only be used in functional modules (FB). Therefore, examples 6.2 and 6.3 are programmed in a functional module (FB), as can be FB1.

In chapter 7 it is explained that every time a different module from the program module PB1 is programmed, it is necessary to program the organization module OB1 which calls the other modules. Therefore, in these two examples the organization module OB1 includes an unconditional calling instruction JU FB1 to the functional module FB1. When the PLC starts the program execution, the program executes organization module OB1, jumps automatically to the functional module FB1 and executes the control program.

EXAMPLE 6.2

Implement, using the Simatic S5-100U PLC, the transition graph or state diagram of the cart control system described in example 1.3.

Solution

Figure 6.27 shows again the transition graph or state diagram of the cart control system. The assignment of input and output variables to the PLC variables is given in Table 6.10. One hot encoding is used to assign to the internal states E_1, E_2 and E_3 the internal variables given in Table 6.11.

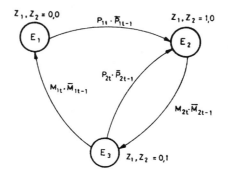

FIGURE 6.27 *Transition graph or state diagram of the control system of example 6.2.*

TABLE 6.10

External variable	SIMATIC S5-100U variable assignment
P_1	I 0.0
P_2	I 0.1
M_1	I 0.2
M_2	I 0.3
Z_1	Q 1.0
Z_2	Q 1.1

TABLE 6.11

Internal state	SIMATIC S5-100U internal state variable		
	F 0.0	F 0.1	F 0.2
E_1	1	0	0
E_2	0	1	0
E_3	0	0	1

TABLE 6.12

Hexadecimal address	Instruction list		Comment
014	: A	F 0.7	⎫
016	: JC	= M0	
018	: AN	F 0.7	⎬ Initial state set up
01A	: =	F 0.0	
01C	: =	F 0.7	⎭
M0	: A	F 0.0	⎫
020	: A	I 0.0	⎬ $P_1 \uparrow$
022	: AN	F 0.3	⎭
024	: =	F 1.0	
026	: AN	F 1.0	
028	: JC	= M1	
02A	: A	F 1.0	⎬ $E_1 \rightarrow E_2$
02C	: =	F 0.1	
02E	: =	Q 1.0	
030	: AN	F 1.0	
032	: =	F 0.0	
M1	: A	F 0.1	⎫
036	: A	I 0.3	⎬ $M_2 \uparrow$
038	: AN	F 0.6	⎭
03A	: =	F 1.0	
03C	: AN	F 1.0	
03E	: JC	= M2	
040	: A	F 1.0	⎬ $E_2 \rightarrow E_3$
042	: =	F 0.2	
044	: =	Q 1.1	
046	: AN	F 1.0	
048	: =	Q 1.0	
04A	: =	F 0.1	
M2	: A	F 0.2	⎫
04E	: A	I 0.1	⎬ $P_2 \uparrow$
050	: AN	F 0.4	⎭
052	: =	F 1.0	
054	: AN	F 1.0	
056	: JC	= M3	
058	: A	F 1.0	⎬ $E_3 \rightarrow E_2$
05A	: =	F 0.1	
05C	: =	Q 1.0	
05E	: AN	F 1.0	
060	: =	Q 1.1	
062	: =	F 0.2	
M3	: A	F 0.2	⎫
066	: A	I 0.2	⎬ $M_1 \uparrow$
068	: AN	F 0.5	⎭
06A	: =	F 1.0	
06C	: AN	F 1.0	
06E	: JC	= M4	
070	: A	F 1.0	⎬ $E_3 \rightarrow E_1$
072	: =	F 0.0	
074	: AN	F 1.0	
076	: =	Q 1.1	
078	: =	F 0.2	
M4	: A	I 0.0	⎬ $P_{1t} \rightarrow P_{1t-1}$
07C	: =	F 0.3	
07E	: A	I 0.3	⎬ $M_{2t} \rightarrow M_{2t-1}$
080	: =	F 0.6	
082	: A	I 0.1	⎬ $P_{2t} \rightarrow P_{2t-1}$
084	: =	F 0.4	
086	: A	I 0.2	⎬ $M_{1t} \rightarrow M_{1t-1}$
088	: =	F 0.5	
08A	: BE		

Internal variables F 0.3, F 0.4, F 0.5 and F 0.6 are used to detect the edges of the input variables P_1, P_2, M_1 and M_2, respectively. For the initial state set-up, internal variable F 0.7 is used. This variable is reset, together with the other internal variables, when the power supply is switched on.

Table 6.12 shows the program in the instruction list language. Decisions to change from one internal state to another are implemented using the conditional jump instruction JC.

We analyze now the instructions executing one of the capacity transitions, for example the capacity transition to change from state E_1 to E_2. First, instruction A F 0.0 in address M0 verifies if the system is at internal state E_1 and those in addresses 020 and 022 verify if a rising edge has been applied to the input variable P_1. Next instruction = F 1.0 stores the result of the preceding operation in the internal variable F 1.0, and the result of the execution of instruction AN F 1.0, is a logic 1 if the content of F 1.0 is a 0. When JC = M1 is executed the PLC makes the following decision:

1. If the state of F 1.0 internal variable is a 0, it jumps to the address M1 to implement the transition capacity from state E_2 to E_3.

2. If the state of F 1.0 is a 1, the PLC executes the following instructions:
   - instructions placed in location 02A and 02C which set the internal state variable F 0.1;
   - instruction placed in location 02E switching on the output variable Q 1.0;
   - instructions placed in locations 030 and 032 which reset the internal state variable F 0.0.

EXAMPLE 6.3

Design a program making the Simatic S5-100U PLC execute the bar selection control system described in example 1.4.

Solution

Figure 6.28 repeats the transition graph or state diagram obtained in example 1.4 from the problem statement. From this state diagram the program is designed systematically using the method described in section 4.5.2.

The assignment of input and output variables to the PLC variables is shown in Table 6.13. One hot encoding is done, assigning variables F 0.0 and F 0.1 to the

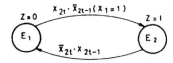

FIGURE 6.28 *Transition graph or state diagram of the control system of example 6.3.*

COMMERCIALLY AVAILABLE PLCS

TABLE 6.13

External variable	SIMATIC S5-100U variable assignment
X_1	I 0.0
X_2	I 0.1
Z	Q 1.0

internal states E_1 and E_2, respectively. Variable F 0.2 is used to detect the edges of X2.

To set the initial conditions, internal variable F 0.3 is used. This variable is reset, like all the other internal variables, when the PLC power supply is switched on.

Table 6.14 shows the program in instruction list, and its analysis is recommended to the reader. The decisions to go from one internal state to another are carried out using the conditional jump instruction JC.

TABLE 6.14

```
Hexadecimal
  address      Instruction list                         Comment

    014        : A        F 0.3      ⎫
    016        : JC       = M0       ⎪
    018        : AN       F 0.3      ⎬   Initial state set up
    01A        : =        F 0.0      ⎪
    01C        : =        F 0.3      ⎭

    M0         : A        F 0.0      ⎫
    020        : A        I 0.1      ⎬ $x_2 \uparrow (x_1 = 1)$
    022        : AN       F 0.2      ⎭
    024        : A        I 0.0      ⎫
    026        : =        F 1.0      ⎪
    028        : AN       F 1.0      ⎪
    02A        : JC       = M1       ⎬   $E_1 \rightarrow E_2$
    02C        : =        F 0.0      ⎪
    02E        : A        F 1.0      ⎪
    030        : =        F 0.1      ⎪
    032        : =        Q 1.0      ⎭

    M1         : A        F 0.1      ⎫
    036        : AN       I 0.1      ⎬ $x_2 \downarrow$
    038        : A        F 0.2      ⎭
    03A        : =        F 1.0      ⎫
    03C        : AN       F 1.0      ⎪
    03E        : JC       = M2       ⎪
    040        : =        F 0.1      ⎬   $E_2 \rightarrow E_3$
    042        : A        F 1.0      ⎪
    044        : =        F 0.0      ⎪
    046        : AN       F 1.0      ⎪
    048        : =        Q 1.0      ⎭

    M2         : A        I 0.1      ⎫
    04C        : =        F 0.2      ⎬   $X_{2t} \rightarrow X_{2t-1}$
    04E        : BE                  ⎭
```

Implementation of level-characterized sequential control systems using the Simatic S5-100U

In this section the same control systems as in section 5.6.3 are executed.

EXAMPLE 6.4

Implement, using the Simatic S5-100U PLC, the automatic control electronic system of the garage described in example 5.3.

TABLE 6.15

External variable	SIMATIC S5-100U variable assignment
S1	I 0.0
S2	I 0.1
S3	I 0.2
S4	I 0.3
S5	I 0.4
M	I 0.5
P	I 0.6
R	I 0.7
M1	Q 2.0
M2	Q 2.1
GL	Q 2.2
RL	Q 2.3

TABLE 6.16

Internal state variable of chapter 5 PLC	Internal state variable of SIMATIC S5-100U
IR0	F 0.5
IR1	F 0.0
IR2	F 0.1
IR3	F 0.2
IR4	F 0.3
IR5	F 0.4

Solution

First we have to assign the PLC's input and output variables to the corresponding external variables. Such assignment is shown in Table 6.15.

The program is designed, like the one of example 5.3, following the methodology of section 5.6.3.

The corresponding program in ladder diagram is represented in Figure 6.29 and is equivalent to that of Figure 5.63. In order for the reader to compare them, Table 6.16 shows the equivalence of the internal state variables of Figures 5.63 and 6.29. Note

```
Segment   1          0000
!
!I 0.5     I 0.6                                                              F 0.5
+---] [---+---]/[---+----------+----------+----------+----------+----------+--(    )-!
!         !
!F 0.5    !
+---] [---+
!
Segment   2          0007
!
!I 0.0     F 0.1     F 0.5     Q 2.3                                          Q 2.0
+---] [---+---]/[---+---] [---+---]/[---+----------+----------+----------+--(    )-!
!         !
!Q 2.0    !
+---] [---+
!
Segment   3          0010
!
!I 0.1     F 0.1     Q 2.0                                                    F 0.0
+---] [---+---]/[---+---] [---+----------+----------+----------+----------+--(    )-!
!         !
!F 0.0    !
+---] [---+
!
Segment   4          0018
!
!I 0.1     F 0.0                                                              F 0.1
+---]/[---+---] [---+----------+----------+----------+----------+----------+--(    )-!
!
Segment   5          001C
!
!I 0.4     F 0.4     F 0.5                                                    F 0.2
+---] [---+---]/[---+---] [---+----------+----------+----------+----------+--(    )-!
!         !
!F 0.2    !
+---] [---+
!
Segment   6          0024
!
!F 0.2     I 0.2     F 0.4                                                    Q 2.1
+---] [---+---] [---+---]/[---+----------+----------+----------+----------+--(    )-!
!         !
!Q 2.1    !
+---] [---+---------+
!
Segment   7          002C
!
!I 0.3     Q 2.1     F 0.4     F 0.5                                          F 0.3
+---] [---+---] [---+---]/[---+---] [---+----------+----------+----------+--(    )-!
!         !
!F 0.3    !
+---] [---+---------+
!
Segment   8          0035
!
!I 0.3     F 0.3                                                              F 0.4
+---]/[---+---] [---+----------+----------+----------+----------+----------+--(    )-!
!
```

FIGURE 6.29 *Program of the control system of the garage of Figure 5.51 using the ladder diagram language of Simatic S5-100U PLC (continued overleaf).*

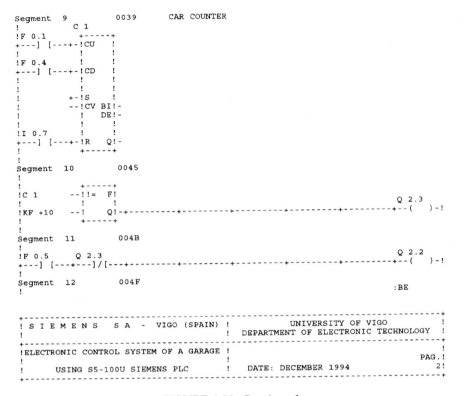

FIGURE 6.29 *Continued.*

that segment 9 of Figure 6.29 shows a reversible counter with a 'set to 0' input and that segment 10 compares the content of the counter with number 10 and activates variable Q 2.3 if the two numbers are the same.

Table 6.17 and Figure 6.30 show the program in the instruction list and the function diagram languages, respectively, of the Simatic S5-100U.

EXAMPLE 6.5

Implement, using the Simatic S5-100U, the electronic control system of the car-washing machine of example 5.4.

Solution

Table 6.18 shows the input and output variable assignment. Figure 6.31 shows the program in the ladder language using the design method orientated towards the internal state variables described in section 5.6.3. This figure is equivalent to

TABLE 6.17

```
Segment  1           0000
    :A(
    :O    I    0.5
    :O    F    0.5
    :)
    :AN   I    0.6
    :=    F    0.5                    START-STOP R-S FLIP-FLOP
    :***


Segment  2           0007
    :A(
    :O    I    0.0
    :O    Q    2.0
    :)
    :AN   F    0.1
    :A    F    0.5
    :AN   Q    2.3
    :=    Q    2.0                    ENTRANCE BARRIER MOTOR CONTROL
    :***


Segment  3           0010
    :A(
    :O    I    0.1
    :O    F    0.0
    :)
    :AN   F    0.1
    :A    Q    2.0
    :=    F    0.0                    "S1" RISE EDGE PULSE GENERATION
    :***


Segment  4           0018
    :AN   I    0.1
    :A    F    0.0
    :=    F    0.1                    "S1" FALL EDGE PULSE GENERATION
    :***


Segment  5           001C
    :A(
    :O    I    0.4
    :O    F    0.2
    :)
    :AN   F    0.4
    :A    F    0.5
    :=    F    0.2                    "S5" RISE EDGE PULSE GENERATION
    :***


Segment  6           0024
    :A(
    :A    F    0.2
    :A    I    0.2
    :O    Q    2.1
    :)
    :AN   F    0.4
    :=    Q    2.1                    EXIT BARRIER MOTOR CONTROL
    :***


Segment  7           002C
    :A(
    :A    I    0.3
    :A    Q    2.1
    :O    F    0.3
    :)
    :AN   F    0.4
    :A    F    0.5
    :=    F    0.3                    "S4" RISE EDGE PULSE GENERATION
    :***
```

```
+------------------------------------------------+------------------------------------------+
! S I E M E N S    S A  -  VIGO (SPAIN)  !        UNIVERSITY OF VIGO                !
!                                         !  DEPARTMENT OF ELECTRONIC TECHNOLOGY    !
+------------------------------------------------+------------------------------------------+
!ELECTRONIC CONTROL SYSTEM OF A GARAGE   !                                          !
!                                         !                                    PAG.!
!       USING S5-100U SIEMENS PLC        !  DATE: DECEMBER 1994                  1!
+------------------------------------------------+------------------------------------------+
```

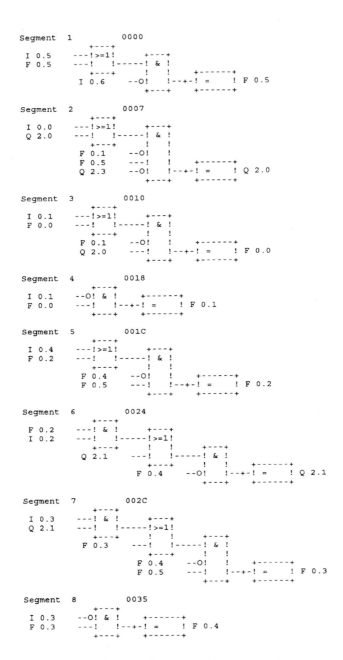

FIGURE 6.30 *Program of the control system of the garage of Figure 5.51 using the function diagram language of Simatic S5-100U PLC.*

```
Segment   9          0039      CAR COUNTER
                C 1
                +-----+
F 0.1        --!CU   !
F 0.4        --!CD   !
             --!S    !
             --!CV BI!-
             !   DE!-
I 0.7        --!R   Q!-
                +-----+

Segment  10          0045
                +-----+
C 1          --!!=  F!
             !      !  +------+
KF +10       --!   Q!-+-! =   ! Q 2.3
                +-----+  +------+

Segment  11          004B
                +---+
F 0.5        ---! & !
Q 2.3        --O!   !-+-! =   ! Q 2.2
                +---+  +------+

Segment  12          004F
                          :BE

+---------------------------------------------------------------+
! S I E M E N S   S A  -  VIGO (SPAIN) !    UNIVERSITY OF VIGO                !
!                                      ! DEPARTMENT OF ELECTRONIC TECHNOLOGY  !
+---------------------------------------------------------------+
!ELECTRONIC CONTROL SYSTEM OF A GARAGE !                                      !
!                                      !                                 PAG.!
!     USING S5-100U SIEMENS PLC        ! DATE: DECEMBER 1994               2 !
+---------------------------------------------------------------+
```

FIGURE 6.30 *Continued.*

TABLE 6.18

External variable	SIMATIC S5-100U variable assignment
S1	I 0.0
S2	I 0.1
S3	I 0.2
M	I 0.3
P	I 0.4
MP1	Q 2.0
MP2	Q 2.1
MV	Q 2.2
MC	Q 2.3
XV	Q 2.4

```
Segment  1            0000
!
!I 0.3     I 0.1      I 0.2      I 0.4      F 0.4                                      F 0.0
+---] [---+---] [---+---] [---+---]/[---+---]/[---+-----------+-----------+--(   )-!
!                    !
!F 0.0               !
+---] [---+---------+
!
Segment  2            000A
!
!F 0.1     F 0.0      F 0.3                                                             Q 2.0
+---]/[---+---] [---+---]/[---+-----------+-----------+-----------+-----------+--(   )-!
!                    !
!F 0.2               !
+---] [---+
!
Segment  3            0012
!
!F 0.0     F 0.2                                                                        Q 2.3
+---] [---+---]/[---+-----------+-----------+-----------+-----------+-----------+--(   )-!
!                                                                                       !
!                                                                                      !Q 2.4
!                                                                                      +--(   )-!
!
Segment  4            0017
!
!I 0.0     I 0.1      F 0.0                                                             F 0.1
+---] [---+---]/[---+---] [---] [---+-----------+-----------+-----------+--(   )-!
!                    !
!F 0.1               !
+---] [---+---------+
!
Segment  5            001F
!
!F 0.1     I 0.0      I 0.1      F 0.0                                                  F 0.2
+---] [---+---]/[---+---] [---+---] [---+---+-----------+-----------+--(   )-!
!                                !                                                      !
!F 0.2                           !                                                     !Q 2.2
+---] [---+---------+---------+                                                        +--(   )-!
!
Segment  6            0029
!
!F 0.1     F 0.2                                                                        Q 2.1
+---] [---+---]/[---+-----------+-----------+-----------+-----------+-----------+--(   )-!
!                    !
!F 0.3     F 0.4     !
+---] [---+---]/[---+
!                    !
!F 0.5               !
+---] [---+---------+
!
Segment  7            0031
!
!F 0.2     I 0.1      I 0.0      F 0.0                                                  F 0.3
+---] [---+---]/[---+---] [---+---] [---+---+-----------+-----------+--(   )-!
!                                !
!F 0.3                           !
+---] [---+---------+---------+
!
Segment  8            003A
!
!F 0.3     I 0.0      I 0.1                                                             F 0.4
+---] [---+---]/[---+---] [---+---+-----------+-----------+-----------+--(   )-!
!
Segment  9            003F
!
!I 0.4     I 0.1      F 0.0                                                             F 0.5
+---] [---+---]/[---+---]/[---+-----------+-----------+-----------+--(   )-!
!          !
!F 0.5     !
+---] [---+
!
Segment  10           0047
!                                                                                      :BE


+-----------------------------------------------------+---------------------------------+
! S I E M E N S    S A  -  VIGO (SPAIN) !      UNIVERSITY OF VIGO                      !
!                                                     ! DEPARTMENT OF ELECTRONIC TECHNOLOGY !
+-----------------------------------------------------+---------------------------------+
!        ELECTRONIC CONTROL SYSTEM       !                                              !
!        OF A CAR WASHING MACHINE        !                                          PAG.!
!        USING S5-100U SIEMENS PLC       ! DATE: DECEMBER 1994                        1!
+-----------------------------------------------------+---------------------------------+
```

FIGURE 6.31 *Program of the control system of the car-washing machine of Figure 5.65 using the ladder diagram language of Simatic S5-100U PLC.*

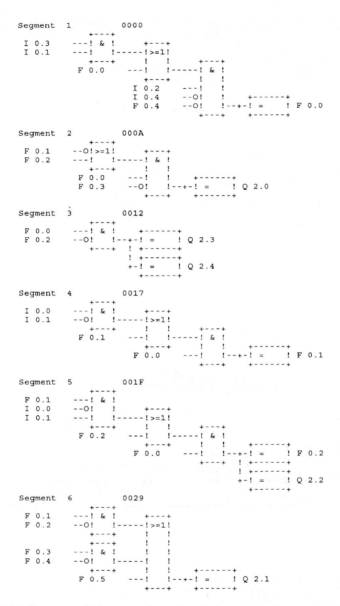

FIGURE 6.32 *Program of the control system of the car-washing machine of Figure 5.65 using function diagram language of Simatic S5-100U PLC (continued overleaf).*

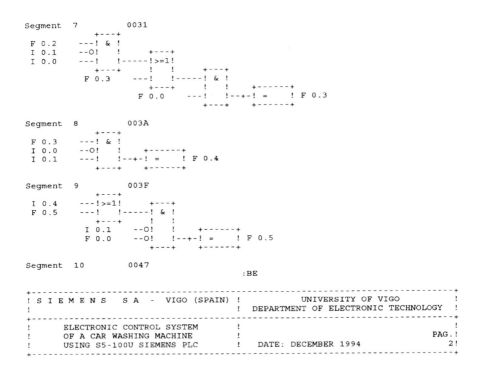

FIGURE 6.32 *Continued.*

TABLE 6.19

Internal state variable of chapter 5 PLC	Internal state variable of SIMATIC S5-100U
IR0	F 0.0
IR1	F 0.1
IR2	F 0.2
IR3	F 0.3
IR4	F 0.4
IR5	F 0.5

TABLE 6.20

```
Segment  1           0000
    :A(
    :A     I    0.3
    :A     I    0.1
    :O     F    0.0
    :)
    :A     I    0.2
    :AN    I    0.4
    :AN    F    0.4
    :=     F    0.0       START-STOP R-S FLIP-FLOP
    :***


Segment  2           000A
    :A(
    :ON    F    0.1
    :O     F    0.2
    :)
    :A     F    0.0
    :AN    F    0.3
    :=     Q    2.0       MP1 MOTOR CONTROL
    :***


Segment  3           0012
    :A     F    0.0
    :AN    F    0.2
    :=     Q    2.3       MOTOR CONTROL OF THE BRUSHES
    :=     Q    2.4       ELECTROVALVE XV CONTROL
    :***


Segment  4           0017
    :A(
    :A     I    0.0
    :AN    I    0.1
    :O     F    0.1
    :)
    :A     F    0.0
    :=     F    0.1       DETECTION OF FIRST TRIP END
    :***


Segment  5           001F
    :A(
    :A     F    0.1
    :AN    I    0.0
    :A     I    0.1
    :O     F    0.2
    :)
    :A     F    0.0
    :=     F    0.2       DETECTION OF SECOND TRIP END
    :=     Q    2.2       FAN CONTROL
    :***


Segment  6           0029
    :A     F    0.1
    :AN    F    0.2
    :O
    :A     F    0.3
    :AN    F    0.4
    :O     F    0.5
    :=     Q    2.1       MP2 MOTOR CONTROL
    :***
```

TABLE 6.20 *Continued*

```
Segment    7              0031
   :A(
   :A    F    0.2
   :AN   I    0.1
   :A    I    0.0
   :O    F    0.3
   :)
   :A    F    0.0
   :=    F    0.3                    DETECTION OF THIRD TRIP END
   :***

Segment    8              003A
   :A    F    0.3
   :AN   I    0.0
   :A    I    0.1
   :=    F    0.4                    DETECTION OF FOUR TRIP END
   :***

Segment    9              003F
   :A(
   :O    I    0.4
   :O    F    0.5
   :)
   :AN   I    0.1
   :AN   F    0.0
   :=    F    0.5                    EMERGENCY STOP
   :***

Segment   10              0047
   :BE
```

```
+---------------------------------------------------------------+
! S I E M E N S   S A  -  VIGO (SPAIN) !      UNIVERSITY OF VIGO              !
!                                      ! DEPARTMENT OF ELECTRONIC TECHNOLOGY  !
+---------------------------------------------------------------+
!      ELECTRONIC CONTROL SYSTEM       !                                  !
!      OF A CAR WASHING MACHINE        !                            PAG. !
!      USING S5-100U SIEMENS PLC       ! DATE: DECEMBER 1994         1!
+---------------------------------------------------------------+
```

Figure 5.66 of example 5.4. To facilitate comparison between them, Table 6.19 indicates the equivalence between the internal state variables of both figures.

Table 6.20 shows the program in the instruction list language of the Simatic S5-100U and Figure 6.32 in function diagram language.

6.3 Sysmac C-28K

Sysmac C-28K is a PLC from Omron with a central unit which can be connected to a programming console. The power supply is external and is offered as a separate module by the manufacturer. All the equipment components can be adapted to be fixed on a rail.

The central unit includes sixteen binary inputs and twelve binary outputs which can be increased by means of external modules. Both the inputs and outputs may be DC or AC. The outputs can be of three types: relay, thyristor or triac.

Each output consists of a standard base block to which any of the modules (relay, thyristor or triac) can be connected. The outputs are grouped in four distinct blocks, and it is possible to use different power supplies for each block.

The central unit has two connectors for the expansion units, programming consoles, memory recording, etc., and one null insertion strength socket (accessible from the outside) for the use of an EPROM memory.

The small programming consoles can be directly connected to the central unit. Figure 6.33 shows this characteristic using model C-20 of the same series. These consoles have a tape connection to store the programs on magnetic tape.

FIGURE 6.33 *Sysmac C-20 PLC with a programming console (courtesy of Omron).*

6.3.1 General characteristics

Some of the characteristics of Sysmac C-28K are:

- Memory capacity: 1194 instructions
- Memory type: RAM and EPROM
- Execution time of each instruction: 10 µs
- Maximum input capacity: 80
- Maximum output capacity: 60
- Internal output variables: 136 (extended to 156)
- Non-volatile internal output variables: 160
- Special internal output variables: 16
- Temporary memory variables: 8
- Timers and counters: 38
    The counters are all non-volatile (they have retention capability)

Timing margin:
  Normal: 0 to 999.99 s
  High speed: 0 to 99.99 s
  Counting margin: 0 to 9999
- Supply of the central unit: 24 V DC
- Battery lifetime: 5 years
- Digital expansion modules: from 16, 24 and 32 I/O variables
- Configuration of the DC digital input modules:
  5–12 V DC/10 mA
  12–24 V DC/10 mA
  12 V DC/7 mA
  24 V DC/7 mA
- Configuration of the AC digital input modules:
  94–121 V AC/10 mA to 100 V AC
  187–242 V AC/10 mA to 200 V AC
  12–24 V AC/10 mA to 24 V AC
  12 V AC/7 mA
  24 V AC/7 mA
- Configuration of the DC digital output modules:
  24 V DC/2 A
  12–48 V DC/300 mA
- Configuration of the AC digital output modules:
  250 V AC/2 A
  85–250 V DC/1 A

6.3.2 Programming

The Sysmac C-28K PLC has three different operating modes selected by a switch situated in the console:

1. Program mode.
2. Monitor mode.
3. Run mode.

The program mode is used to program the PLC step by step using the instruction list language. The run mode is used to execute the program stored in memory, and the monitor mode is used when we want to display the PLC evolution during program execution. In the monitor mode it is possible to observe the evolution of the timers and counters, and also to verify the state of the input and output (external or internal) variables.

The PLC executes the program when it is in the run mode. For execution to occur, the programming console must be disconnected from the PLC while the operation mode switch is in the run position. If this is not the case then the PLC will not execute the program unless a power failure occurs. When power returns, the PLC restarts in the run mode.

This PLC has also the following characteristics:

1. It allows programming (depending on the type of console) in the instruction list and ladder diagram languages. A computer with the appropriate programming resources must be used for programming in the function diagram language.
2. It uses the program block concept, represented by a combination of contacts constituting a multifunction from which several output variables are generated. This concept is described in detail later in this section.

Variable identification

The Sysmac C-28K equipment has the following variable types:

- External I/O variables
- Internal output variables
- Non-volatile internal output variables
- Special internal output variables
- Temporary memory internal variables
- Timers
- Counters

EXTERNAL I/O VARIABLES
To identify this variable type, the PLC uses the concept of channels and dots. Each channel is composed of sixteen dots (sixteen I/O binary variables). Each binary variable is identified by four digits, from which the two on the left indicate the channel number and the two on the right the dot or I/O variable inside the channel. Figure 6.34 shows an example of identification of inputs and external outputs.

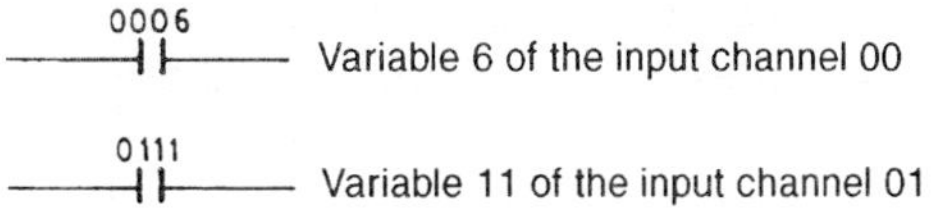

```
        0006
      ──┤ ├──    Variable 6 of the input channel 00

        0111
      ──┤ ├──    Variable 11 of the input channel 01
```

FIGURE 6.34 *Example of input variable identification.*

The basic configuration of this PLC has one input channel (channel 00) and one output channel (channel 01) in the central unit. Its maximum capacity is ten I/O channels; the I/O channels are automatically assigned at the time that expansion modules are added.

Only the first twelve variables of each output channel (0100 to 0111) are available from the outside. The other four (0112 to 0115) may only be used as internal output variables.

INTERNAL OUTPUT VARIABLES

This PLC has 136 internal output variables which are also identified with four digits. These variables are assigned to channels 10 to 17 and half of channel 18, that is from 1000 to 1807. It should be remembered that the last four variables of each external channel may be used as internal output variables, as indicated previously.

NON-VOLATILE INTERNAL OUTPUT VARIABLES

These variables behave like internal output variables (above) except that they are non-volatile, i.e. they keep their value when a power failure occurs. They are grouped in ten channels (from 0 to 9) and are identified by the characters HR followed by three digits, the one on the left indicating the channel number and the two on the right the variable number inside the channel. Therefore, these variables are included between HR000 and HR915.

SPECIAL INTERNAL OUTPUT VARIABLES

This PLC has sixteen internal binary variables which are called special because they are switched on or off according to the operating state of the PLC and do not depend on the state of any input or output variable. These variables are identified by numbers 1808 to 1815 and 1900 to 1907 and have the following function:

- 1808: This variable is switched on when the PLC detects a battery failure.
- 1809: This variable is switched on when the I/O cycle of the PLC is between 100 and 130 ms.
- 1810: This variable is switched on during the first execution cycle following the I/O cycle in which the input 0001 resets the high-speed counter.
- 1811: In normal operation this variable is switched off.
- 1812: In normal operation this variable is switched off.
- 1813: In normal operation this variable is switched on.
- 1814: In normal operation this variable is switched off.
- 1815: This variable is switched on during the first I/O cycle following the run mode set of the PLC.
- 1900: Generates a pulse sequence of period 0.1 s. When it is combined with a counter it acts as a timer and retains its value if a power failure occurs.
- 1901: Identical to the 1900 with a period of 0.2 s.
- 1902: Identical to the 1900 with a period of 1 s.
- 1903-07: These variables are used with the arithmetic instructions described in chapter 7.

Variables 1811 to 1814 can be combined with external variables to indicate the operating mode of the PLC.

COMMERCIALLY AVAILABLE PLCS 247

TEMPORARY MEMORY INTERNAL VARIABLES

These variables are used together with the program block concept which is discussed later in this section. The system has eight variables of this type which are identified by characters TR and a number between 0 and 7. The use of these variables in the same block does not have to be sequential, but one variable can only be used once in a block although it can be used in different blocks.

TIMERS AND COUNTERS

These variables are identified with characters TIM and CNT, respectively, followed by two digits between 00 and 47 indicating the order number of the timer or counter.

As can be seen in the programming section, there are high-speed timers, which are represented by TIMH and a number between 00 and 47 and up/down counters represented by CNTR and a number between 00 and 47. This identification (TIMH or CNTR) is only used in the high-speed timer or up/down counter instruction since the variables they generate are represented by TIMXX and CNTXX.

Instruction set

LOGIC AND OUTPUT INSTRUCTIONS

Load function (LOAD)

This function indicates the first variable of a block or set of variables. It is represented by the code LD followed by the identification number of the variable. Figure 6.35 shows an application example.

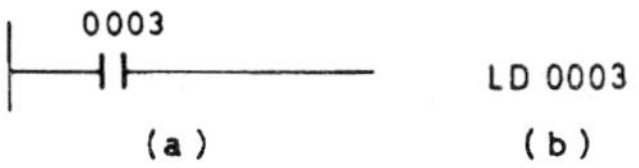

```
        0003
    ──┤ ├──              LD 0003
      (a)                 (b)
```

FIGURE 6.35 *Example of LOAD function.*

Output function (OUT)

This function switches on or off an external or internal output variable depending on the result of the previous logic operation. An example is given in Figure 6.36.

Inversion logic function (NOT)

This function inverts or negates the associated variable. It is combined with all the other functions and is represented by characters NOT. Its use can be seen in Figures 6.37 to 6.54.

```
        0002    0106
    ─┤ ├─────( )─────          LD 0002
                                OUT 0106

          (a)                     (b)
```

FIGURE 6.36 *Example of OUT function.*

```
    0009   HR214   0012    0103
    ─┤ ├───┤ ├─────┤/├────( )──       LD 0009
                                       AND HR214
                                       AND NOT 0012
                                       OUT 0103

              (a)                         (b)
```

FIGURE 6.37 *Example of the logic AND function.*

Logic AND function (AND)
This function executes the logic AND function between the variable specified in the preceding instruction (or the result of the preceding sequence of logic instructions) and the variable associated with it. It is represented by the code AND followed by the identification number of the variable. An example is shown in Figure 6.37.

Logic OR function (OR)
This function executes the logic OR function between the variable specified in the preceding instruction (or the result of the preceding sequence of instructions) and the variable associated with it. It is represented by the code OR followed by the identification number of the variable. An example is shown in Figure 6.38.

Logic AND function of OR functions (AND LD)
This function executes the AND function between groups of OR functions. It is represented by the code AND LD. An example is given in Figure 6.39.

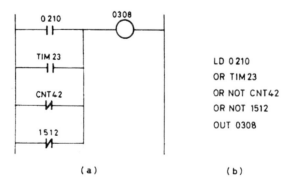

FIGURE 6.38 *Example of the logic OR function.*

COMMERCIALLY AVAILABLE PLCS 249

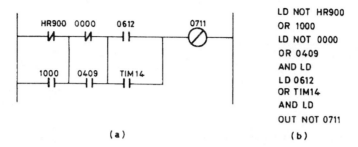

FIGURE 6.39 *Example of the logic AND function of logic OR functions (AND LD)*.

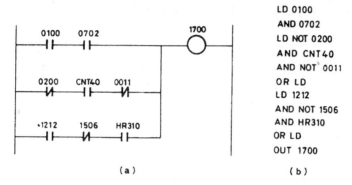

FIGURE 6.40 *Example of the logic OR function of logic AND functions (OR LD)*.

Logic OR function of AND functions (OR LD)
This function executes the OR function between groups of AND functions. It is represented by the code OR LD. An example is given in Figure 6.40.

Figures 6.41 and 6.42 show two examples of logic functions combining all the previous instructions.

TIMING, COUNTING AND S–R (OR SELF-RETENTION) INSTRUCTIONS

Timer (TIM)
To program a timer we use one variable or a logic combination of variables preceding the TIM instruction. The timer output is switched on when the preceding result is a logic 1 and is switched off when the result is a logic 0.

The timer instruction occupies only one memory location including the timer number (00 to 47) and the preselecting time (0 to 999.9). The timer constitutes a binary variable and can switch on an output variable directly.

Figure 6.43 indicates a 10 second timer operating as a delay to switching on and Figure 6.44 a 10 second timer operating as a delay to switching off.

250   PROGRAMMABLE LOGIC CONTROLLERS

```
LD 0200
AND NOT 1302
LD CNT38
AND NOT 0001
OR LD
LD TIM19
OR 0511
AND LD
LD HR005
AND 1807
AND 0712
LD 1310
AND NOT HR405
OR LD
AND LD
OUT HR617
```

(a)   (b)

FIGURE 6.41 *Example of combination of different logic functions including OR LD and AND LD, and the output function.*

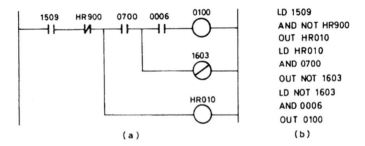

```
LD 1509
AND NOT HR900
OUT HR010
LD HR010
AND 0700
OUT NOT 1603
LD NOT 1603
AND 0006
OUT 0100
```

(a)   (b)

FIGURE 6.42 *Example of a multifunction combining different logic functions.*

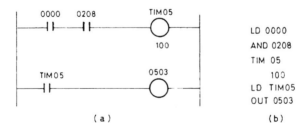

```
LD 0000
AND 0208
TIM 05
    100
LD TIM05
OUT 0503
```

(a)   (b)

FIGURE 6.43 *Example of a switch-on delay timer.*

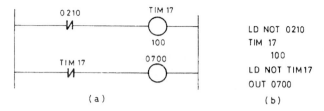

FIGURE 6.44 *Example of a switch-off delay timer.*

High-speed timer (TIMH)
The only difference between this and the previously described timer is the selection margin which is of 00.00–99.99 seconds with increments of 0.01 seconds in this case. Figure 6.45 shows a 200 millisecond timer.

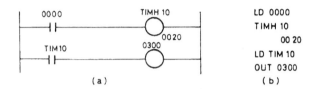

FIGURE 6.45 *Example of a high-speed timer.*

Counter (CNT)
To program a counter, two variables or sequences of variables are needed: one for the counting input (CP) and another for the reset input (R). The counting input acts in the rising edge of the pulses to be counted. The reset input acts when its value is logic 1 and allows the evolution of the counter when its value is logic 0. If the two inputs are applied simultaneously, the reset input overrides the counting input.

The CNT instruction only occupies one memory location, including the counter number (00 to 47) and the preselected value (0 to 9999).

All the counters retain their contents when a power failure occurs. This counter operates as a down-counter. Figure 6.46 shows an application example and Figure 6.47 the corresponding timing diagram.

Up/down counter (CNTR)
This counter is different from CNT (above) because it has three inputs: up-counting (ACP), down-counting (SCP) and reset (R).

If the up- and down-counting pulses are applied simultaneously, the contents of the counter do not change. An application example is represented in Figure 6.48, and Figure 6.49 shows the corresponding timing diagram.

Although the instruction is called CNTR, the variable acting over the output variable is identified as CNT. This counter counts up when the preselection value is overtaken, and counts down when the preselection value

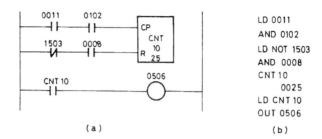

FIGURE 6.46 *Example of a counter.*

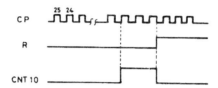

FIGURE 6.47 *Timing diagram of the variables of the counter of Figure 6.46.*

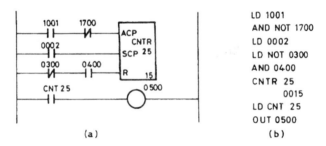

FIGURE 6.48 *Example of an up/down counter.*

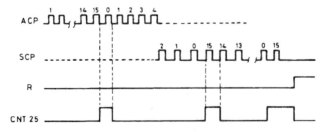

FIGURE 6.49 *Timing diagram of the variables of the counter of Figure 6.48.*

is reached. In both cases it stays active until the following counting pulse (ascending in the first case and descending in the second).

S–R instruction (KEEP)

This instruction behaves like an S–R flip-flop. It has two inputs: set and reset. A variable, which can be external, internal or non-volatile internal output, is associated with the KEEP instruction. If this variable is a non-volatile internal output its value is kept when a power failure occurs.

When programming this instruction the first variable used is the set and the second is the reset.

Figure 6.50 shows an application example and Figure 6.51 its corresponding timing diagram.

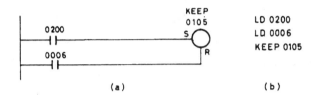

FIGURE 6.50 *Example of the S–R instruction (KEEP) application.*

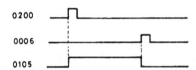

FIGURE 6.51 *Timing diagram of the variables of the example of Figure 6.50.*

CONTROL INSTRUCTIONS

In chapter 5 it is shown that control instructions provide decision-making capabilities by changing the instruction execution sequence according to the results of arithmetic and logic instructions. The control instructions of Sysmac C-28K are similar to those described in chapter 5 and are characterized by their simplicity. The types of control instruction are:

JMP–JME

These two instructions are combined and have the same function as the JMP–JME instructions described in section 5.2.2.

IL–ILC

These two instructions are also combined. All the output variables included between IL and ILC are switched off if the result of the logic instruction or sequence of logic instructions placed immediately before IL is 0. If timers and counters exist between IL and ILC, the first are also switched off and the second keep their value. If, on the other hand, the result of the logic instruction or sequence of logic instructions placed before IL is 1, the PLC executes all the instructions between IL and ILC as if they were not included in the program (Figure 6.52).

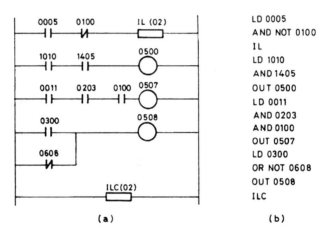

(a)   (b)

FIGURE 6.52 *Example of IL and ILC control instructions application.*

PROGRAM BLOCK AND ASSOCIATED VARIABLES

As indicated at the beginning of this section, the Sysmac C-28K PLC uses the concept of a program block, this being a multifunctional block with several output variables. Figure 6.53 shows the ladder diagram of a program block, which is associated with the concept of temporary memory internal variable TR, and Figure 6.54 shows an application example in the ladder diagram and the instruction list languages.

FIGURE 6.53 *Example of a program block using the ladder diagram language.*

COMMERCIALLY AVAILABLE PLCS   255

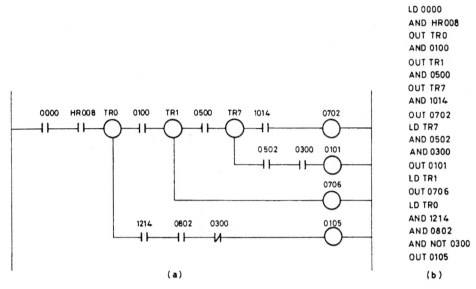

FIGURE 6.54 *Example of a program block using (a) the ladder diagram and (b) the instruction list languages.*

6.3.3 Practical examples of digital system implementation using the Sysmac C-28K

Some of the control systems described in chapters 1 and 5 are now designed using the Sysmac C28K PLC.

Combinational system implementation using the Sysmac C-28K

EXAMPLE 6.6

Design a program for the Sysmac C-28K PLC using the instruction list, ladder diagram and function diagram languages to implement the supervising system of the process of example 3.1.

Solution
The equations are the same as in example 6.1:

$$PP = LSL + TSL + DPSH + PSL + BS1 \cdot BS2$$
$$XV1 = \overline{BS1}$$
$$XV2 = \overline{BS2}$$
$$XV3 = PP + PSH$$
$$GL = \overline{PP}$$
$$RL = PP$$

Table 6.21 indicates the assignment of the input and output variables to the PLC variables. The program in the instruction list language is shown in Table 6.22. Figures 6.55 and 6.56 show the program in the ladder diagram and function diagram languages, respectively.

TABLE 6.21

External variable	SYSMAC C-28K variable assignment
LSL	0000
TSL	0001
DPSH	0002
PSL	0003
PSH	0004
BS1	0005
BS2	0006
PP	0100
XV1	0101
XV2	0102
XV3	0103
GL	0104
RL	0105

TABLE 6.22

```
Instruction list              Comment

    LD      0000     ⎫
    OR      0001     ⎪
    OR      0002     ⎪
    OR      0003     ⎬  PP variable generation
    LD      0005     ⎪
    AND     0006     ⎪
    OR        LD     ⎪
    OUT     0100     ⎭

    LD NOT  0005     ⎫
    OUT     0101     ⎬  XV1 variable generation

    LD NOT  0006     ⎫
    OUT     0102     ⎬  XV2 variable generation

    LD      0100     ⎫
    OR      0004     ⎬  XV3 variable generation
    OUT     0103     ⎭

    LD NOT  0100     ⎫
    OUT     0104     ⎬  GL variable generation

    LD      0100     ⎫
    OUT     0105     ⎬  RL variable generation
```

COMMERCIALLY AVAILABLE PLCS   257

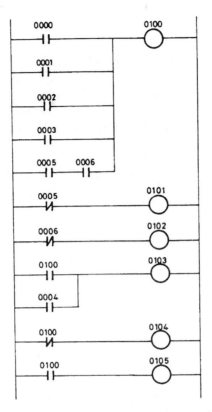

FIGURE 6.55 *Program of the supervising system of example 6.6 using the ladder diagram language.*

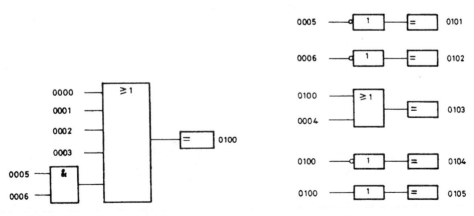

FIGURE 6.56 *Program of the supervising system of example 6.6 using the function diagram language.*

Implementation of edge-characterized sequential control systems using the Sysmac C-28K

In the synthesis of asynchronous sequential systems characterized by edges the PLC must execute control instructions to decide if actions that produce a change of internal state are to be executed or not. In this section two control systems implemented with the Sysmac C-28K PLC are analyzed. Both systems use the JMP and JME instructions described in section 6.3.2.

In the following two examples it is unnecessary to specify the address of each instruction since jump instructions are not used.

EXAMPLE 6.7

Design a program using the Sysmac C-28K PLC to implement the state diagram of Figure 5.40 corresponding to the control of the cart described in example 1.3. (The graph is repeated as Figure 6.57.)

TABLE 6.23

External variable	SYSMAC C-28K variable assignment
P_1	0000
P_2	0001
M_1	0002
M_2	0003
Z_1	0100
Z_2	0101

TABLE 6.24

Internal state	SYSMAC C-28K internal state variable		
	1000	1001	1002
E_1	1	0	0
E_2	0	1	0
E_3	0	0	1

TABLE 6.25

Instruction list			Comment
LD NOT	1007		
JMP			
OUT	1000		Initial state set up
OUT	1007		
JME			
LD	1000		
AND	0000	} P_1 ↑	
AND NOT	1003		
JMP			$E_1 \rightarrow E_2$
OUT NOT	1000		
OUT	1001		
OUT	0100		
JME			
LD	1001		
AND	0003	} M_2 ↑	
AND NOT	1006		
JMP			$E_2 \rightarrow E_3$
OUT NOT	1001		
OUT	1002		
OUT NOT	0100		
OUT	0101		
JME			
LD	1002		
AND	0001	} P_2 ↑	
AND NOT	1004		
JMP			$E_3 \rightarrow E_2$
OUT NOT	1002		
OUT	1001		
OUT NOT	0101		
OUT	0100		
JME			
LD	1002		
AND	0002	} M_1 ↑	
AND NOT	1005		
JMP			$E_3 \rightarrow E_1$
OUT NOT	1002		
OUT	1000		
OUT NOT	0101		
JME			
LD	0000		$P_{1t} \rightarrow P_{1t-1}$
OUT	1003		
LD	0001		$M_{2t} \rightarrow M_{2t-1}$
OUT	1004		
LD	0002		$P_{2t} \rightarrow P_{2t-1}$
OUT	1005		
LD	0003		$M_{1t} \rightarrow M_{1t-1}$
OUT	1006		
END			

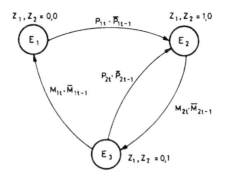

FIGURE 6.57 *Transition graph or state diagram of the control system of example 6.5.*

Solution
The output and input variables are assigned as in Table 6.23. One hot encoding is carried out and the internal variables listed in Table 6.24 are assigned to the internal states E_1, E_2 and E_3.

To detect the edges of P_1, P_2, M_1 and M_2, internal variables 1003, 1004, 1005 and 1006 are used. Internal variable 1007 is used for initial state set-up.

Table 6.25 shows the program in the instruction list language.

EXAMPLE 6.8

Design a program using the Sysmac C-28K PLC to implement the bar selection control system described in example 3.4.

Solution
Figure 6.58 repeats the state diagram obtained in example 3.4 from the problem statement. The program is designed using the same method as in example 6.3.

The assignment of input and output variables is shown in Table 6.26, and Table 6.27 shows the assignment of internal states to the internal state variables of the PLC.

To detect the edges of variable X_2, internal variable 1002 is used. Internal variable 1003 is used for initial state set-up.

Table 6.28 shows the corresponding instruction list program.

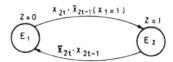

FIGURE 6.58 *Transition graph or state diagram of the control system of example 6.8.*

COMMERCIALLY AVAILABLE PLCS

TABLE 6.26

External variable	SYSMAC C-28K variable assignment
X_1	0000
X_2	0001
Z	0100

TABLE 6.27

Internal state	SYSMAC C-28K internal state variable	
	1000	1001
E_1	1	0
E_2	0	1

TABLE 6.28

```
Instruction list                                Comment

LD NOT   1003      ⎫
JMP                ⎪
OUT      1000      ⎬   Initial state set up
OUT      1003      ⎪
JME                ⎭


LD       1000      ⎫
AND      0001      ⎪
AND NOT  1002      ⎬  x₂ ↑ (x₁ = 1)
AND      0000      ⎪
JMP                ⎪
OUT NOT  1000      ⎬   E₁ → E₂
OUT      1001      ⎪
OUT      0100      ⎪
JME                ⎭


LD       1001      ⎫
AND NOT  0001      ⎬  x₂ ↓
AND      1002      ⎪
JMP                ⎪
OUT NOT  1001      ⎬   E₂ → E₃
OUT      1000      ⎪
OUT NOT  0100      ⎪
JME                ⎭


LD       0001      ⎫
OUT      1002      ⎬   x₂ₜ → x₂ₜ₋₁
```

Implementation of level-characterized sequential control systems using the Sysmac C-28K

The same control systems as described in section 5.6.3 are now designed using the Sysmac C-28K PLC.

EXAMPLE 6.9

Implement, using the Sysmac C-28K PLC, the electronic control system of the garage described in example 5.3.

Solution
We first assign the input and output variables of the PLC to the corresponding external variables (Table 6.29).

TABLE 6.29

External variable	SYSMAC C-28K variable assignment
S1	0000
S2	0001
S3	0002
S4	0003
S5	0004
M	0005
P	0006
R	0007
M1	0100
M2	0101
GL	0102
RL	0103

The program is designed, like the one of example 5.3, using the design method orientated towards the output variables which is described in section 5.6.3.

The corresponding program in the ladder diagram language is shown in Figure 6.59 which is equivalent to that of Figure 5.63. To facilitate the comparison between them, Table 6.30 shows the equivalence between internal state variables of Figures 5.63 and 6.59.

A decimal number together with the letters PRM is associated with each contact line of Figure 6.59. This division corresponds to a division of the program in

COMMERCIALLY AVAILABLE PLCS

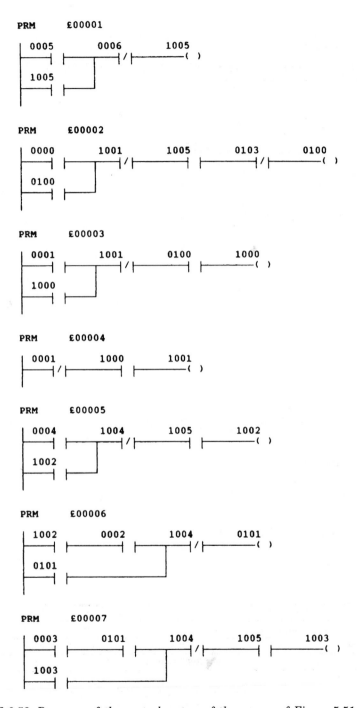

FIGURE 6.59 *Program of the control system of the garage of Figure 5.51 using the ladder diagram language of Sysmac C-28K PLC (continued overleaf).*

264   PROGRAMMABLE LOGIC CONTROLLERS

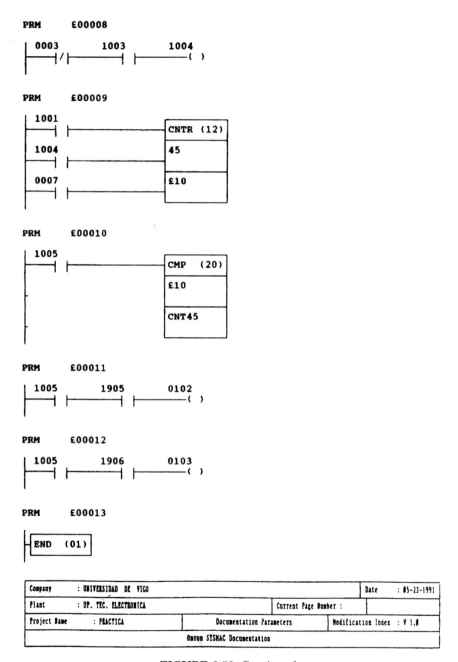

FIGURE 6.59 Continued.

TABLE 6.30

Internal state variable of chapter 5 PLC	Internal state variable of SYSMAC C-28K
IR0	1005
IR1	1000
IR2	1001
IR3	1002
IR4	1003
IR5	1004

TABLE 6.31

```
PRM          £00001

LD           0005
OR           1005
AND NOT      0006
OUT          1005
NETWORK

PRM          £00002

LD           0000
OR           0100
AND NOT      1001
AND          1005
AND NOT      0103
OUT          0100
NETWORK

PRM          £00003

LD           0001
OR           1000
AND NOT      1001
AND          0100
OUT          1000
NETWORK

PRM          £00004

LD NOT       0001
AND          1000
OUT          1001
NETWORK

PRM          £00005

LD           0004
OR           1002
AND NOT      1004
AND          1005
OUT          1002
NETWORK
```

TABLE 6.31 Continued

```
PRM         £00006

LD          1002
AND         0002
OR          0101
AND NOT     1004
OUT         0101
NETWORK


PRM         £00007

LD          0003
AND         0101
OR          1003
AND NOT     1004
AND         1005
OUT         1003
NETWORK

PRM         £00008

LD NOT      0003
AND         1003
OUT         1004
NETWORK


PRM         £00009

LD          1001
LD          1004
LD          0007
CNTR (12)   45            £10
NETWORK

PRM         £00010

LD          1005
CMP (20)    £10           CNT45
NETWORK

PRM         £00011

LD          1005
AND         1905
OUT         0102
NETWORK

PRM         £00012

LD          1005
AND         1906
OUT         0103
NETWORK

PRM         £00013

END (01)
NETWORK
```

Company	: UNIVERSIDAD DE VIGO			Date	: 05-23-1991
Plant	: DP. TEC. ELECTRONICA		Current Page Number :		
Project Name	: PRACTICA	Documentation Parameters		Modification Index	: V 1.0
	Omron SISMAC Documentation				

COMMERCIALLY AVAILABLE PLCS

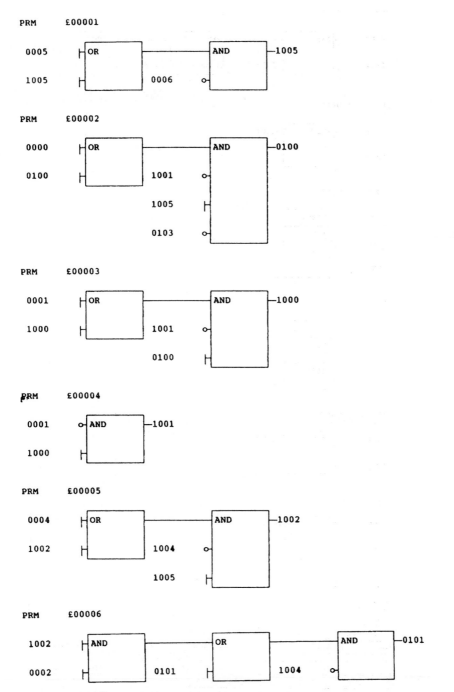

FIGURE 6.60 *Program of the control system of the garage of Figure 5.51 using the function diagram language of Sysmac C-28K PLC (continued overleaf).*

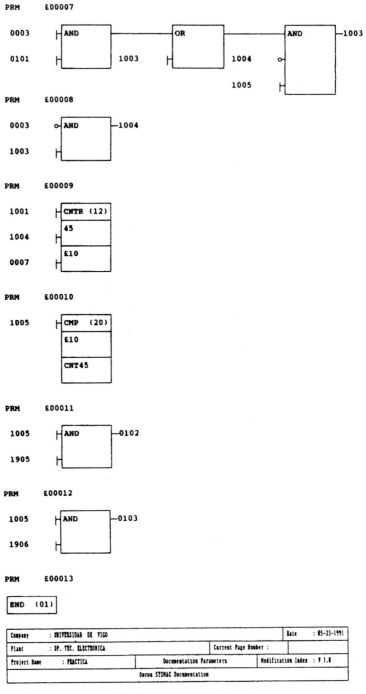

FIGURE 6.60 *Continued.*

segments. Note segments PRM £ 00009 and PRM £ 00010 of the program. The first one corresponds to an up/down counter (CNTR 45) which counts cars and operates according to the description set out in section 6.3.2. Segment PRM £ 00010 compares (CMP) the contents of the car counter CNTR 45 with number 10 (described in section 7.4.2). If the result of the comparison indicates that number 10 is greater than the counter content, special internal variable 1905 is switched on, and if they are the same, special internal variable 1906 is switched on (see section 6.3.2).

Table 6.31 shows the program in the instruction list language and Figure 6.60 in the function diagram language of the Sysmac C-28K PLC. Each segment of Table 6.31 begins with the letters PRM followed by £ XXXXX (where X is a decimal digit), and ends with the word NETWORK.

EXAMPLE 6.10

Implement, using the Sysmac C-28K PLC, the electronic control system of the car-washing machine of example 5.4.

Solution

Table 6.32 gives the assignment of input and output variables to the PLC variables.

Figure 6.61 shows the program in the ladder diagram language of the Sysmac C-28K, implemented using the design method orientated towards the internal state variables described in section 5.6.3.

Figure 6.61 is equivalent to Figure 5.66 of example 5.4. To facilitate their comparison, Table 6.33 indicates the equivalence between the internal state variables of both figures.

Table 6.34 and Figure 6.62 show the program in the instruction list and the function diagram languages, respectively, of the Sysmac C-28K PLC.

TABLE 6.32

External variable	SYSMAC C-28K variable assignment
S1	0000
S2	0001
S3	0002
M	0003
P	0004
MP1	0100
MP2	0101
MV	0102
MC	0103
XV	0104

```
PRM    £00001
 ├─┤0003─┤ ├0001─┬┤ ├0002──┤ ├0004──┤/├1004──┤/├1004────(1000)
 ├─┤1000─┘
```

```
PRM    £00002
 ├─┤/├1001─┬─┤ ├1000──┤/├1003──(0100)
 ├─┤ ├1002─┘
```

```
PRM    £00003
 ├─┤ ├1000──┤ ├1002──┬┤/├0103──(0103)
                     └──(0104)
```

```
PRM    £00004
 ├─┤ ├0000─┬┤/├0001──┤ ├1000──(1001)
 ├─┤ ├1001─┘
```

```
PRM    £00005
 ├─┤ ├1001─┬┤/├0000──┤ ├0001──┤ ├1000─┬─(1002)
 ├─┤ ├1002─┘                          └─(0102)
```

```
PRM    £00006
 ├─┤ ├1001──┤/├1002─┬─(1001)
 ├─┤ ├1003──┤/├1004─┘
 ├─┤ ├1005
```

FIGURE 6.61 *Program of the control system of the car-washing machine of Figure 5.65 using the ladder diagram language of Sysmac C-28K PLC.*

COMMERCIALLY AVAILABLE PLCS

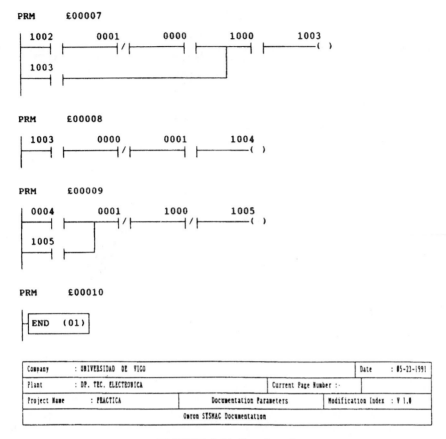

FIGURE 6.61 Continued.

TABLE 6.33

Internal state variable of chapter 5 PLC	Internal state variable of SYSMAC C-28K
IR0	1000
IR1	1001
IR2	1002
IR3	1003
IR4	1004
IR5	1005

TABLE 6.34

```
PRM        £00001

LD         0003
AND        0001
OR         1000
AND        0002
AND NOT    0004
AND NOT    1004
OUT        1000
NETWORK

PRM        £00002

LD NOT     1001
OR         1002
AND        1000
AND NOT    1003
OUT        0100
NETWORK

PRM        £00003

LD         1000
AND NOT    1002
OUT        0103
OUT        0104
NETWORK

PRM        £00004

LD         0000
AND NOT    0001
OR         1001
AND        1000
OUT        1001
NETWORK

PRM        £00005

LD         1001
AND NOT    0000
AND        0001
OR         1002
AND        1000
OUT        1002
OUT        0102
NETWORK

PRM        £00006

LD         1001
AND NOT    1002
LD         1003
AND NOT    1004
OR LD
LD         1005
OR LD
OUT        1001
NETWORK
```

TABLE 6.34 *Continued*

```
PRM       £00007

LD        1002
AND NOT   0001
AND       0000
OR        1003
AND       1000
OUT       1003
NETWORK

PRM       £00008

LD        1003
AND NOT   0000
AND       0001
OUT       1004
NETWORK

PRM       £00009

LD        0004
OR        1005
AND NOT   0001
AND NOT   1000
OUT       1005
NETWORK

PRM       £00010

END  (01)
NETWORK
```

Company	: UNIVERSIDAD DE VIGO			Date	: 05-23-1991
Plant	: DP. TEC. ELECTRONICA		Current Page Number : .		
Project Name	: PRACTICA	Documentation Parameters		Modification Index	: V 1.0
	Omron SYSMAC Documentation				

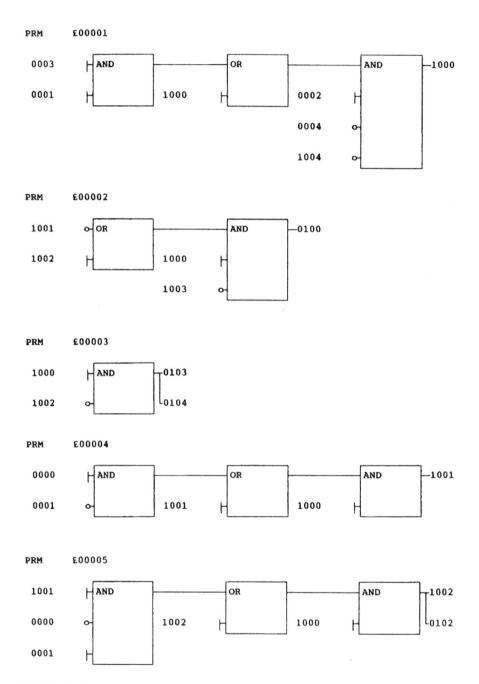

FIGURE 6.62 *Program of the control system of the car-washing machine of Figure 5.65 using the function diagram language of Sysmac C-28K PLC.*

COMMERCIALLY AVAILABLE PLCS

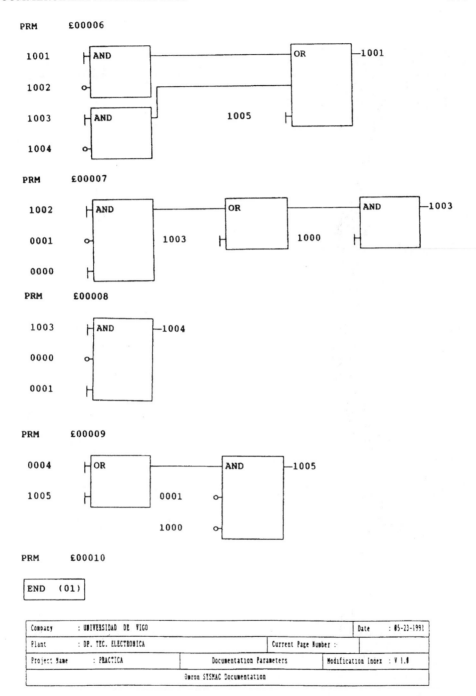

FIGURE 6.62 *Continued.*

CHAPTER 7

Microprocessor-based PLCs

7.1 Introduction

In chapters 4 and 5 it is shown that PLCs are able to execute binary logic operations and that the program is executed sequentially. Sometimes, however, the electronic control system of an industrial process needs to be able to tackle various other situations. For example:

1. Many industrial processes have analog variables which, when converted into digital, give rise to a set of binary variables representing numerical data. Therefore, the electronic control system needs to have numerical processing capabilities.
2. Sometimes the electronic system has to react immediately to input variable changes.

It is time now to consider the following questions. What is the difference between a basic PLC and a computer? Are they substitutes for each other? To answer them, it is necessary to analyze the evolution of PLCs since their inception at the end of the 1960s.

The first PLCs were a consequence of the development of the transistor–transistor logic (TTL) bipolar technology and facilitated the use of electronic control systems in industrial processes where it was impossible to use a computer owing to its high cost. The first PLCs were implemented in accordance with the theory described in chapter 4 and from their analysis and comparison with a computer, some conclusions may already be drawn:

1. A basic PLC only executes logic operations with independent binary variables. On the other hand, a computer can execute the same operations and also arithmetic or logic operations with combinations of a given number of bits (8, 16, 32, etc.).
2. A basic PLC looks at all the input and output variables by executing instructions in a unique sequence. For example, the change of an input

variable is not detected until the PLC arrives at the corresponding selection instruction. This is because the control unit has only one counter with no external interrupt facilities.

The control unit of a computer, on the other hand, has two counters. One evolves in a periodic sequence searching for instructions in memory locations and executing them through the generation of a set of control signals. The other changes its contents (by increasing or replacing them by the address contained in the instruction) every time the first counter executes one cycle. The contents of this second counter (called 'program counter') constitute the address used by the first counter to search for the instruction. Therefore, if the first counter observes some specific external signals before initiating a new cycle, the computer is able to execute immediately a special instruction sequence designed to perform a given task. The computer is then said to have **interrupt capacity**. In spite of this, basic PLCs are able to solve numerous industrial control problems, and, in fact, they replaced the automatic systems formerly implemented with relays because of their superior performance.

Has this situation changed during the 1970s and the 1980s? The answer is undoubtedly yes. The developments in microelectronics stemming from the increase in integration capacity brought about the introduction, first in several integrated circuits (1971) and later in a single one (1974), of the CPU (central processing unit) which united the control and the arithmetic units of a computer. The microprocessor was born, leading to a substantial reduction in computer costs.

As a consequence of all these developments, PLC manufacturers have seen the practical merits of producing PLCs with microprocessors and endowing them with:

- languages orientated towards the resolution of control problems, such as the examples studied in chapter 5, and
- the ability to process analog variables as well as numerical and alphanumerical digital information.

By the end of the 1970s, PLCs with numerical and alphanumerical information processing capability implemented via a microprocessor began to appear on the market.

We can therefore define actual PLCs as computers with resources orientated towards the solution of industrial control problems.

Before studying PLCs with numerical processing capabilities, it is worth reviewing the basic concepts of microprocessors.

7.2 Microprocessors and control applications

An approach analogous to that studied in chapter 4 for the transformation of a synchronous sequential system into a PLC can be adopted for its transformation into a computer. The following are required:

1. An instruction memory.
2. An arithmetic and logic unit.
3. A control unit with the ability to perform fetch and execute cycles.

The block diagram of a computer is shown in Figure 7.1. The control unit has a simplified state diagram with only two distinct states (Figure 7.2a):

1. A fetch state where the pulses required to read the contents of one location of the instruction memory are generated.
2. An execute state where the pulses required to execute the instruction are generated.

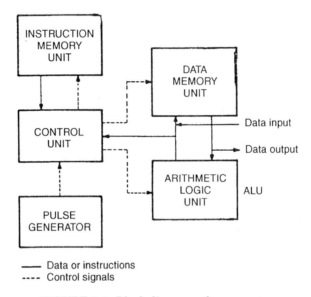

—— Data or instructions
---- Control signals

FIGURE 7.1 *Block diagram of a computer.*

Figure 7.2b shows the basic block diagram of a computer control unit. The reader must verify that it is far more complex than its counterpart of a basic PLC.

The format of a computer instruction is similar to that of a PLC (Figure 7.3) and consists of a combination of bits indicating to the control unit the actions it should execute. This means that instructions can be stored in the same memory as the data thereby reducing the number of control unit terminals. Consequently we obtain at the block diagram of Figure 7.4 which is the one adopted in the majority of computers. The most appropriate structure for the memory unit is random access since the fetching of data or instructions requires the same number of generator pulses regardless of where the items are located in the memory.

MICROPROCESSOR-BASED PLCS

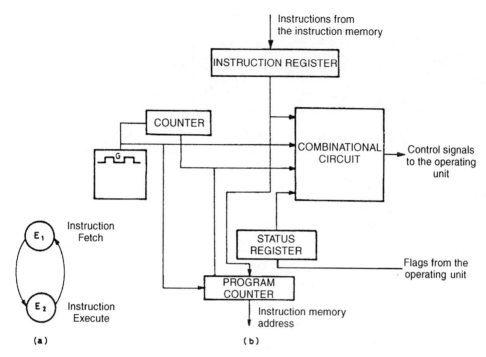

FIGURE 7.2 *Control unit of a computer.* (a) *Simplified state diagram.* (b) *Block diagram.*

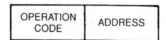

FIGURE 7.3 *Computer instruction format.*

On the other hand, the external data and the partial results must be stored in a read/write random access or active memory (normally called RAM). The instructions can also be stored in a RAM but normally they do not have to be modified during operation and a passive memory in any of its versions (ROM, PROM, EPROM, EEPROM or FLASH) can be used. Semiconductor passive memories have the great advantage of being non-volatile and are therefore reliable where the storage of program instructions is concerned. When the instructions are stored in a passive memory, the diagram of Figure 7.4 becomes that of Figure 7.5. The instruction memory and the data memory are physically separated but communicate with the control unit via the same connections.

From what has been said we conclude that the control unit directs and the arithmetic unit executes the process. Hence, both blocks can be condensed into a single one which receives the name CPU (central processing unit), as

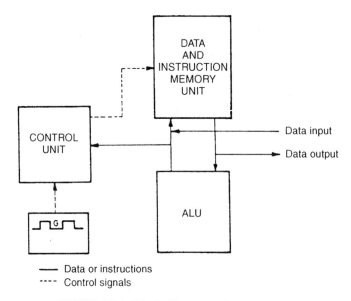

FIGURE 7.4 *Block diagram of a computer.*

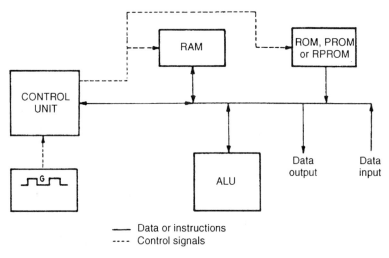

FIGURE 7.5 *Block diagram of a computer using RAM and ROM, PROM or RPROM (EPROM, EEPROM, or FLASH) memories.*

indicated in Figure 7.6. The block diagram of a computer then becomes that of Figure 7.7. The connection between the two blocks consists of:

- a set of control signals that the CPU sends to the memory, and
- a set of binary variables through which the CPU sends information to the memory or vice versa.

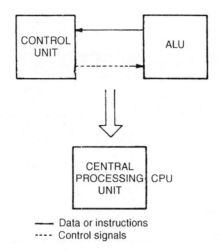

FIGURE 7.6 *Central processing unit (CPU)*.

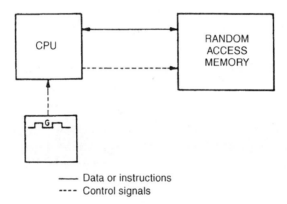

FIGURE 7.7 *Block diagram of a computer*.

In general, the data to be processed have to be transferred from the outside world into the computer and the results have to be transferred in the opposite direction. In addition, the instructions are often stored in an external memory and have to be sent to the random access memory linked to the CPU.

The external systems are known as **peripherals** and normally the computer exchanges information with just one of them at a time. Therefore, the typical structure of a computer becomes that shown in Figure 7.8.

Progress in integration techniques enabled the CPU of a computer to be condensed into a single integrated circuit, under the name of **microprocessor**. The computer whose CPU is a microprocessor is known as **microcomputer**.

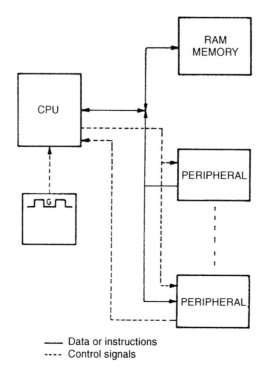

FIGURE 7.8 *Typical architecture of a computer.*

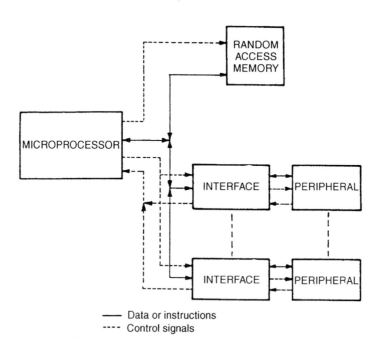

FIGURE 7.9 *Block diagram of a microcomputer.*

On the other hand, most of the peripherals are synchronous sequential digital systems with a different clock from the CPU, hence they cannot be connected with the microprocessor directly. In these cases it is necessary to include a synchronization coupling unit (normally called 'interface') with a memory whose organization (RAM, FIFO, CAM, etc.) depends on the characteristics of the peripheral and on the program (software) located in the computer. In the simplest cases, the interface is just a parallel input/output register.

Hence we arrive at the block diagram of a microcomputer shown in Figure 7.9. If the different peripherals are condensed into one block and the same for all the interface units, the block diagram of Figure 7.10 is obtained, where the memory unit is, in turn, divided into two blocks: active memory (RAM) and passive memory (ROM, PROM, EPROM, EEPROM or FLASH).

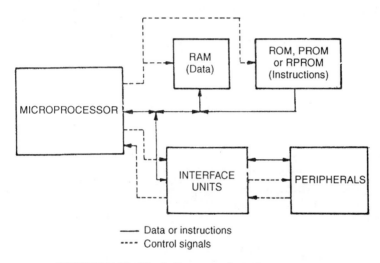

FIGURE 7.10 *Block diagram of a microcomputer.*

It must be emphasized that interface units are functional blocks whose correct design is vital in order to be sure that operation specification changes affect only the instructions stored in the memory.

7.3 PLCs implemented using a microprocessor

7.3.1 General characteristics

A microprocessor can have as interfaces the input and output modules studied in section 1.3.2 of chapter 1. If the memory contains an appropriate instruction sequence, it is evident that a microcomputer can perform the same tasks as a basic PLC. Therefore, a computer implemented with a microprocessor can behave as a PLC and execute a control program not only with digital input and output variables but also with analog ones. To such characteristics we can add the possibility of incorporating communication interfaces.

We then obtain the typical block diagram of a microprocessor-based PLC shown in Figure 7.11. This system is identical to the block diagram of a microcomputer in Figure 7.10 and has the following characteristics:

1. **Program memory**. This is divided into two parts containing:
   - in one part, an executive program or monitor whose function is to execute a set of essential tasks, such as loading the control program coming from an external programming unit, loading digital variables during the input/output cycle, etc., and others that improve the PLC performance, e.g. checking the operation of all PLC elements;
   - in the other part, a memory where we locate the control program which is normally designed by the user in a programming unit.

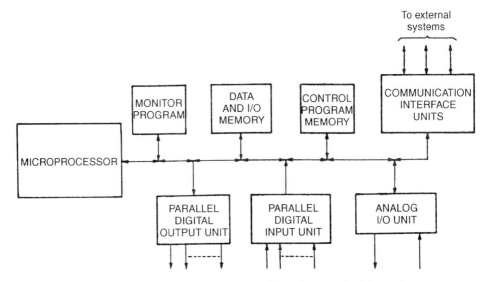

FIGURE 7.11 *Typical block diagram of a PLC implemented with a microprocessor.*

2. **Data memory**. This consists of a RAM divided into the following parts:
   - data or temporal memory of the monitor program;
   - memory of digital input and output variables;
   - memory of numerical data coming from analog-to-digital converters or obtained as operation results which are normally transmitted to digital-to-analog converters;
   - memory of internal state variables.

3. **Interfaces for digital and analog input and output variables**. Digital variables are interfaced via the modules described in section 1.3.2 (Figures 1.32 to 1.38). Analog variables are interfaced via interface units executing analog-to-digital and digital-to-analog conversions.

4. **Communication interfaces**. These allow the PLC to be connected to external electronic systems which perform a variety of functions; they include:
    - programming units to write the control program in a given language and transfer it to the corresponding PLC memory. These are studied in chapter 8;
    - remote digital and analog data input and output units;
    - management computers integrating the data given by the PLC with other data to produce statistics, production graphs, characteristics of the process operation (stopping times, production rates, etc.).

From the above we can conclude that microprocessor-based PLCs have a typical memory map as shown in Figure 7.12.

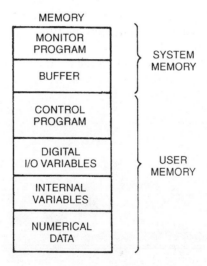

FIGURE 7.12 *Typical memory map of a PLC implemented with a microprocessor.*

7.3.2 Hardware and software resources

From the block diagram of Figure 7.11, we can study the characteristics of each element and its practical implementation.

The combination of elements with different characteristics gives rise to a large variety of PLCs either from the same or from different manufacturers. The most important characteristics are described below.

PROGRAM MEMORY CAPACITY

This is defined as the maximum number of instructions which can be entered into the control program memory using the instruction list language. PLC capacity varies but in general it is between 1 and 64 Kbytes.

Memory capacity is related to input/output capacity, discussed below. The greater the maximum number of input and output variables of a PLC, the greater should be its memory capacity.

DIGITAL INPUT AND OUTPUT CAPACITY

This is an important and significant characteristic. It is defined as the maximum number of digital input and output variables that can be connected to the PLC and can be used to classify PLCs as:

- I/O capacity less than 128
- I/O capacity between 128 and 512
- I/O capacity greater than 512

There are many variants including:

1. A central unit containing a given and fixed number of inputs and outputs which can be increased with expansion modules. An example is the Sysmac C-28K PLC from Omron (Figure 7.13).
2. A central unit with neither I/O capacity nor input or output modules. Examples are S5-115U and S5-150 PLCs from Siemens (Figure 7.14) and Sysmac C-200 PLC from Omron (Figure 7.15).

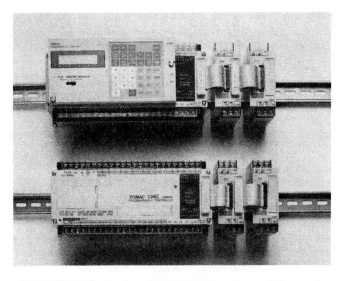

FIGURE 7.13 *Sysmac C-28K PLC (courtesy of Omron)*.

MICROPROCESSOR-BASED PLCS   287

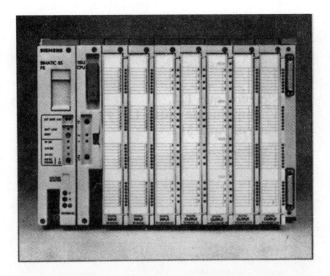

FIGURE 7.14 *S5-115U and S5-150 PLCs (courtesy of Siemens).*

Another important parameter is the programming language. This parameter is related to the existence of external programming units which are connected to the PLC via the relevant communication interface unit.

The use of a microprocessor in the central unit allows the control programs to be organized as in a computer, with a main program module (organization module) and functional modules constituting subroutines. Numerical operations, e.g. magnitude comparisons, additions, subtractions, multiplications, divisions, etc., are also made possible by a microprocessor.

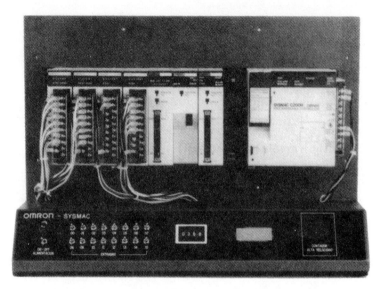

FIGURE 7.15 *Sysmac C-200H PLC (courtesy of Omron)*.

7.4 Examples of PLCs with numerical processing capabilities

The PLCs studied in chapter 6 are microprocessor-based and therefore constitute an example of PLCs with arithmetic capabilities. In that chapter their binary logic capability was studied; in this section, we go further and analyze their arithmetic and analog signal processing characteristics.

7.4.1 Simatic S5-100U

Hardware

ANALOG SIGNAL PROCESSING
The Simatic S5-100U PLC has analog input modules executing the conversion of analog variables to digital ones (A/D) and analog output modules converting digital variables to analog ones (D/A). The maximum number of analog inputs and outputs is eight in total.

The general characteristics of the analog input modules are:

- Several configurations:
    4 inputs of ±50 mV
    4 inputs of ±1 V

    4 inputs of ±10 V
    4 inputs of ±20 mA
    4 inputs of 4 to 20 mA
    2 inputs of ±500 mV
- Connection of the module with the signal emitter: two wires
- Digital input information: 11 bits plus sign bit, equivalent to 2048 decimal units
- A/D conversion time: 60 ms at 50 Hz
- A/D conversion method: double ramp integration
- Integration time: 20 ms at 50 Hz
- Selection switches for:
    network frequency: 50 or 60 Hz
    number of required inputs: 1, 2 or 4
    wire breakage signal (in the modules of ±50 mV, ±1 V and ±500 mV)

The general characteristics of the analog output modules are:

- Several configurations:
    2 outputs of ±10 V
    2 outputs of ±20 mA
    2 outputs of 4 to 20 mA
    2 outputs of 1 to 5 V
- Digital output information: 10 bits plus sign bit, equivalent to 1024 decimal units
- Conversion time: 0.1 ms
- Load connection:
    using two wires in current loop configurations
    using four wires in voltage loop configurations, two of which are a high impedance sensor line (S+ and S−) to prevent the output signal being disturbed by the voltage drop across the wires

ACCUMULATORS AND OPERANDS

This equipment has two 16-bit accumulators ACCU1 and ACCU2 and it is possible to select:

- Output combinations:
    bytes QB
    words (16 bit) QW
- Input combinations:
    bytes IB
    words (16 bits) IW
- Data words (16 bits) divided into:
    left byte DL
    right byte DR

- Marks or internal outputs:
    mark byte FB
    mark word (16 bits) FW
- Constants:
    one byte
    two alphanumerical characters KC
    16-bit fixed point numbers (between -32768 and $+32767$) KF
    hexadecimal (from 0 to FFFF) KH
    counting (from 0 to 999) KC
    timing KT

Software

Because the Simatic S5-100U is a microprocessor-based PLC, it has instructions for the processing of binary combinations and allows the programs to be organized into software modules which are related using structured programming techniques.

TRANSFER AND ARITHMETIC INSTRUCTIONS
The most important instructions are analyzed below.

Transfer and load instructions

There are two load instructions, identified by characters L and LC. Both load the accumulator ACCU1 with the new data and transfer its former contents to ACCU2.

The L instruction is used to load one of the following:

- One input word (IW)
- One output word (QW)
- One mark word (FW)
- One data word (DW)
- One timing value (KT)
- One counting value (KC)
- One fixed point constant (KF)
- One hexadecimal constant (KH)

The LC instruction is used when the information loaded into ACCU1 is BCD coded.

The transfer instruction is identified with character T and transfers data contained in ACCU1 to a certain word (input, output or mark). For instance, the transference of input byte 2 to output word 17 is implemented using instructions:

        L IB 2
        T QW 17

The T QW instruction also stores the digital information applied to the analog output modules into the digital I/O memory creating a process image of the outputs.

Arithmetic instructions

These instructions execute the addition or subtraction of the contents of ACCU1 and ACCU2. The add instruction is identified by +F and the subtraction by −F. The latter executes the subtraction (ACCU2) − (ACCU1).

The instruction sequence below executes the subtraction of fixed point constant (+825) and the contents in the data word 12, and stores the result in the output word 28:

    L KF +825
    L DW 12
    −F
    T QW 28

The result of the arithmetic operations is flagged in one of three ways:

- CC0: This is a logic 1 when the result is between −32768 and −1 or when it is greater than +32767.
- CC1: This is a logic 1 when the result is between +1 and +32767 or when it is less than −32768.
- OV: This is a logic 1 when the result is less than −32768 or greater than +32767.

Magnitude comparison instruction

This instruction compares the contents of ACCU1 and ACCU2. The different comparisons which can be programmed, together with their identifying characters, are listed in Table 7.1.

The comparisons are executed between the contents of ACCU2 and ACCU1. Flags CC1 and CC0 are both 0 if (ACCU1) = (ACCU2); 0 and 1, respectively, if (ACCU2) < (ACCU1); and 1 and 0, respectively, otherwise.

TABLE 7.1

Comparison identification	Comparison operation
I = F	Equal to detection
> < F	Not equal to detection
> F	Greater than detection
> = F	Greater than or equal to detection
< F	Less than detection
< = F	Less than or equal to detection

The instruction sequence indicated below detects whether input word 7 is greater than or equal to input word 23 and activates output A 1.2 if that is the case:

```
L IW 7
L IW 23
> = F
= Q 1.2
```

Complementary operating instructions
These instructions can only be used in the functional modules. They allow the execution of three types of operation:

- Combinational operations using words
- Transformation operations
- Shift operations

Combinational operations with words. There are three instructions that execute logic operations with the contents of accumulators ACCU1 and ACCU2. These logic operations are done bit by bit between the contents of the two accumulators and the result is stored in accumulator ACCU1.

The instructions executing these operations are:

- AW
- OW
- XOW

The AW instruction executes the logic AND operation between the two accumulator contents; OW executes the logic OR, and XOW the exclusive OR.

The following sequence executes the exclusive OR function between input words 7 and 23 and stores the result in output word 14:

```
L IW 7
L IW 23
XOW
T QW 14
```

Transformation operations. There are two instructions acting upon accumulator ACCU1 contents:

- CFW
- CSW

Instruction CFW executes the one's complement of the accumulator contents and instruction CSW executes the two's complement. In both cases the result is stored in ACCU1.

The following sequence executes the two's complement of the input word 15 and stores the result in mark word 60:

    L IW 15
    CSW
    T FW 60

Shift operations. There are two instructions to shift left or right the data in accumulator ACCU1. Each instruction also indicates the number n of positions being shifted (n is a number between 0 and 15). The two instructions executing these operations are:

- SLW n
- SRW n

The SLW n instruction shifts the word in ACCU1 n positions to the left, and the SRW n instruction executes the same operation to the right. In both cases the result is stored in ACCU1.

The following sequence shifts data word 15 two positions to the right and stores the result in output word 20:

    L DW 15
    SRW
    T QW 20

Block transfer instruction

This instruction is identified with characters TNB and transfers data from one internal memory block to another. To define the magnitude of the blocks, a number is associated with this instruction indicating the number of bytes to be transferred.

The programming of this transfer is done by loading into accumulator ACCU2 the final address of the block being transferred and into accumulator ACCU1 the final address of the block in the new destination. Because the PLC memory addresses are expressed in hexadecimal, their loading is done with operand KH.

The sequence below executes the transfer of a block of five bytes between addresses 01A4 and 01A8 (both inclusive) to the memory locations between addresses 1273 and 1277:

    LKH 01A8
    LKH 1277
    TNB 5

SPECIAL INSTRUCTIONS

We include in this section the following instructions:

- STP
- NOP

The STP instruction changes the PLC internal state from run to stop if the conditions of the process under control require it.

In the sequence below, the PLC stops the execution of the program if input variable I 0.7 is switched off:

    A I 0.7
    JC PB10
    STP

The NOP instruction executes a null operation, that is it allows memory positions to be left empty. This facilitates the automatic translation of the program from the instruction list to the ladder diagram or function diagram languages.

JUMP TO A MODULE AND RETURN INSTRUCTIONS

There are three different instructions to execute a module:

- JU
- JC
- CDB

The JU instruction is used to execute an unconditional call to a module (program or functional), that is a call is executed regardless of the result of the logic operation before JU.

The JC instruction is used to execute a conditional call to a module (program or functional), that is a call is executed only when the result of the operation before JC is a 1.

The instructions JU and JC include the identification of the module they call.

The following sequence shows program modules PB8 and PB10 calling. PB10 is executed only if input I 0.3 is a 1:

    A I 2.4
    A F 0.2
    =Q 1.5
    JC PB8
    A I 0.3
    JC PB10

The ADB instruction is used to execute unconditional calls of data modules.

A data module call is always unconditional and requires the CDB instruction followed by the module identification number.

Return instructions end the execution of a module and can be conditional or unconditional. There are three return instructions:

- BE
- BEU
- BEC

The BE instruction constitutes an unconditional end of the module. It is not necessary to program it because when an order is given to the programming console to transfer the program to the PLC memory, instruction BE is automatically included at the end of the program.

The BEU instruction constitutes an unconditional return from a module to the point where the module was called. In the following sequence, if Q 1.2 is a logic 1, the functional module FB14 is executed, while if Q 1.2 is a logic 0, then when the BEU instruction is reached the return to module PB2 takes place:

```
JU PB2
JU FB14
...
...
A I 0.1
A I 0.2
= Q 1.2
JC XXX
BEU
...
...
XXX ...
BE
```

The BEC instruction is similar to BEU except that the former is conditional, that is it only ends the module if the result of the previous logic operation is a 1.

STRUCTURED PROGRAMMING

Structured programming consists in dividing the control task into several subtasks and designing a program for each of them. Each program is included in a program module (PB). Finally an organization module (OB) is used to establish the order and processing conditions of each program (PB), functional (FB) or data (DB) module.

Structured programming has the following advantages:

1. Each part of the process can be independently programmed, therefore a useful overview can be gained of the set of machines or processes to be controlled.
2. The different parts of the program can be checked independently, making the start-up task and the isolation and location of faults easier.

Besides the modules referred to above, there are functional (FB) and data modules (DB). Functional modules execute more complex functions than program modules (PB) and may be developed by the user, although they are provided as standard products by Siemens. Data modules (DB) have 256 words each and store the data required for program execution.

The Simatic S5-100U provides the following modules for structured programming development.

Organization modules (OB)

- **OB1**. This module is executed cyclically and establishes the processing hierarchy of the program (PB), functional (FB) and data (DB) modules.
- **OB21 and OB22**. These modules are processed only once before starting the execution of OB1, in the case of structured programming, and before PB1 if it is linear programming. They are used to establish the behaviour of the PLC before starting cyclic processing. Both organization modules OB21 and OB22 may be programmed.
- **OB34**. This module is processed cyclically before OB1 or PB1 and defines the behaviour of the PLC in case of a battery failure. This module can be programmed.

Program modules (PB)

The equipment has 64 (0–63) program modules available. If there is more than one PB, it is necessary to program OB1.

Functional modules (FB)

The equipment also has 64 (0–63) functional modules. A name with a maximum of eight characters can be assigned to each of them.

Data modules (DB)

The equipment has 64 (0–63) data modules to store numerical data in binary, decimal or hexadecimal form or alphanumerical characters.

Both program (PB) and functional (FB) modules can be executed using conditional or unconditional call instructions. When a conditional call is used, module execution depends on the result of the logic operation immediately before the call instruction.

The data modules (DB) are only unconditionally called and they remain valid inside an organization module (OB), a program module (PB) or functional module (FB) until another data module is called.

The organization modules (OB) cannot be called. The PLC executes them or not according to the type of module and the case in hand. From an organization module (OB), a program module (PB) or a functional module (FB), other program (PB) or functional (FB) modules can be called. The equipment has a 16-level nested call capacity. The combination of this set of modules is shown in Figure 7.16.

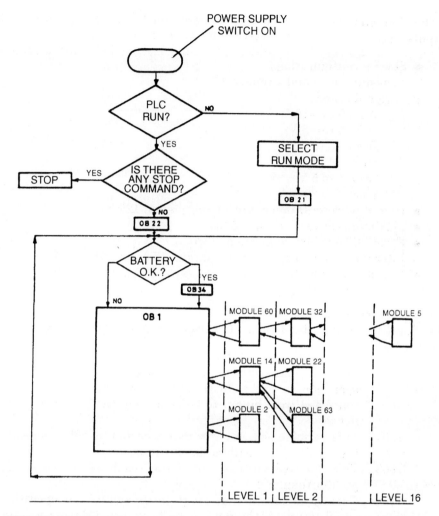

FIGURE 7.16 *Combination of the program modules of Simatic S5-100U PLC.*

7.4.2 Sysmac C-28K

Hardware

ANALOG SIGNAL PROCESSING

The Sysmac C-28K PLC has analog input modules converting analog variables to digital ones and analog output modules converting digital variables to analog ones. The maximum number of analog I/O modules is five inputs and five outputs.

The characteristics of the analog variable modules, for both inputs and outputs, are:

- Several configurations:
    modules of 2 and 4 inputs
    signal ranges:
      from 0 to +10 V
      from 0 to +5 V
      from −5 to +5 V
      from −10 to +10 V
      from 1 to 5 V
      from 4 to 20 mA
- Digital representation of the analog signal: 12 bits
- Maximum conversion time: 5 ms
- External variables connection: 2 wires
- Internal consumption of analog inputs: 250 mA at 5 V DC
- Internal consumption of analog outputs: 250 mA at 5 V DC

Internal binary variables

The flags indicating the result of various numerical operations are as follows:

1903: This variable switches on when the result of an arithmetic operation is not a BCD number or when the data to be converted from binary to BCD or from BCD to binary exceed 9.999.
1904: This variable acts as carry indicator and its value depends on the result of the arithmetic operations.
1905: This variable switches on if the result of an operation is greater than (>).
1906: This variable switches on if the result of an operation is equal to (=), if the data to be transferred during a transfer operation are 0 and if the result of an arithmetic operation is 0.
1907: This variable switches on if the result of an operation is less than (<).

Software

Because it is a microprocessor-based PLC, the Sysmac C-28K has instructions for processing binary combinations. Some of the most important instructions from the point of view of its arithmetic capability are presented below.

TRANSFER AND ARITHMETIC INSTRUCTIONS

Transfer instructions

There are two transfer instructions:

- MOV
- MVN

The MOV instruction (code for MOVE) transfers a four-digit hexadecimal constant or the data from one channel to another. This instruction is associated with one input variable (or combination of variables). Only when the associated variable or combination of variables has logic value 1 does the PLC execute MOV. The first channel to be programmed is the channel of data to be transferred and the second the destination channel.

The MVN instruction (code for MOVE NOT) acts in much the same way as MOV, the difference being that the data channel items are inverted during instruction execution.

When the MOV or MVN instructions are executed, internal variable 1906 switches on if the value of all the bits of the transferred channel or constant (16 bits) is a logic 0.

Figure 7.17 shows two transfer operations (both executed when variables 1201 and HR302 are logic 1). The first operation transmits the four-digit hexadecimal constant 150A to internal output channel 15 and the second inverts the data of channel HR2 and transfers them to the external output channel 05.

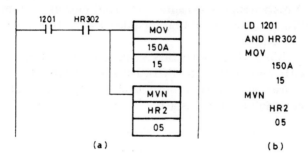

FIGURE 7.17 *Transfer instruction example.*

Set and reset of the carry flag instructions

As indicated in previous sections, the internal variable 1904 acts as a carry flag. This variable, besides being set when a carry is generated after an arithmetic operation, can also be set or reset using STC or CLC instructions.

The STC instruction sets the variable 1904 and the CLC instruction resets it. Both instructions are associated with one binary variable (or a logic function of several variables) and are executed, or not, depending on the

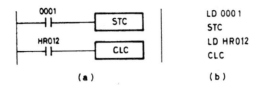

FIGURE 7.18 *Example of the carry flag set and reset instructions.*

logic value of the variable. Figure 7.18 indicates a sequence which sets variable 1904 if the value of variable 0001 is a logic 1, and resets it if the value of variable HR012 is a logic 1.

Arithmetic instructions

There are four arithmetic instructions (addition, subtraction, multiplication and division) and all of them execute in BCD the corresponding operation between two channels or between one channel and a four-digit BCD constant. The result is stored in a third channel.

The numbers used in all the operations should be in BCD code. If any is not, internal output variable 1903 is set and the operation is not executed. In the same way, internal output variable 1906 is set when the result of any of the operations is 0000.

All the arithmetic instructions are associated with a binary variable or combination of binary variables and the operation is executed, or not, depending on whether the logic value of the result is 1 or 0, respectively.

The **addition instruction** is identified by ADD. If a carry is generated when the ADD instruction is executed, internal output variable 1904 is set. Hence, a CLC instruction should be included before the ADD instruction to reset the carry flag 1904.

Figure 7.19 shows the instruction sequence executing the addition of channel 15 with the four-digit BCD constant 8206. The result of the operation is stored in internal output channel HR5. This operation is executed only if variable HR014 is a 1.

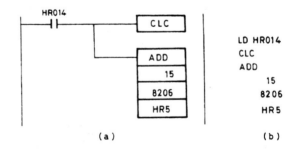

FIGURE 7.19 *Example of addition instruction application.*

The **subtraction instruction** is identified by SUB. If the result of the operation is negative, internal output variable 1904 is set. Therefore, and similarly to the addition instruction, a CLC instruction should be programmed before instruction SUB to reset variable 1904.

The first channel or four-digit BCD constant to be programmed is the minuend and the second the subtrahend. Figure 7.20 indicates the instruction sequence subtracting the contents of internal output channel 12 from 3729 and storing the result in internal output channel 15. The operation is executed if variable 1000 is a 1.

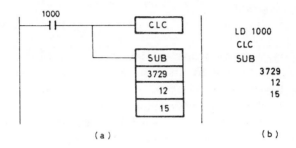

FIGURE 7.20 *Example of subtraction instruction application.*

The **multiplication instruction** is identified by MUL. The first channel or four-digit constant in BCD to be programmed is the multiplicand and the second the multiplier. Figure 7.21 indicates the instruction sequence executing the multiplication of constant 1739 by the contents of internal output channel HR3 and storing the result in internal output channel HR9. The operation is executed when binary variables 0002 and HR015 are logic 1 simultaneously.

The **division instruction** is identified by DIV. The first channel or four-digit constant in BCD to be programmed is the dividend and the second the divisor. Figure 7.22 indicates the instruction sequence executing the division of HR2 by HR5 and storing the result in channel 10 internal output. This operation is executed when one or both binary variables 0013 and 1300 are a logic 1.

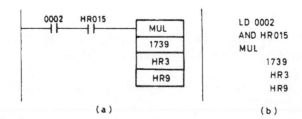

FIGURE 7.21 *Example of multiplication instruction application.*

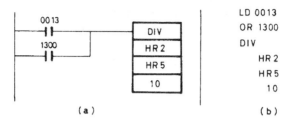

FIGURE 7.22 *Example of division instruction application.*

Code conversion instructions
There are two instructions which convert a binary number to BCD and vice versa.

BCD to binary conversion is identified by BIN. This instruction converts the BCD data of a channel into a 16-bit binary number. The result is stored in an, internal output or external output channel.

To program the BIN instruction we also need a variable or combination of binary variables. The BIN instruction is only executed if the indicated binary variable is a logic 1. This instruction has two associated channels, the first containing the data to be converted and the second being the channel where the result of the conversion is stored.

If the result of the conversion is 0 (all bits equal to zero), internal output variable 1906 is set. If the information of the first channel is not in BCD, internal output variable 1903 is set and the conversion does not take place.

Figure 7.23 shows an instruction sequence that converts the data (in BCD) of channel 10 to binary and stores the result in channel 12. The operation is executed when variables 0000 and HR000 are simultaneously 1.

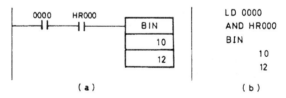

FIGURE 7.23 *Example of natural BCD (usually called BCD) to natural binary (usually called binary) conversion.*

Binary to BCD conversion is identified by BCD. This instruction converts the binary data of a 16-bit channel into its BCD equivalent and stores the result in an, internal output or external output channel.

The BCD instruction is also associated with a binary variable, or combination of binary variables, and is executed only if its logic value is 1. This instruction has two associated channels, one containing the data to be converted and the other receiving the conversion result. If the conversion result is 0 (all bits are equal to zero), internal output variable 1906 is set. Similarly, if the information stored in the destination channel exceeds 9999, the internal output signal 1903 is set and the instruction is not executed.

Figure 7.24 indicates the instruction sequence which converts the data (in binary) of channel 10 to BCD and stores the result in channel 12. The operation is executed when variables 0000 and HR000 are simultaneously a logic 1.

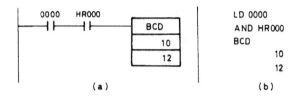

FIGURE 7.24 *Example of binary to BCD conversion.*

Magnitude comparison instruction
This instruction is identified by CMP and compares one 16-bit channel or a constant with other 16-bit channels. The channels can be inputs, external outputs, internal outputs, a timing or counting value, or a hexadecimal constant of any value between 0000 and FFFF.

The CMP instruction is also associated with a binary variable (or combination of binary variables) and is executed only if the value of the associated variable is a logic 1. The result of the comparison is obtained via the special internal outputs 1905, 1906 and 1907, which are set if the result is greater than, equal to or less than respectively.

Figure 7.25 shows an instruction sequence executing the comparison of internal output channel HR3 and the 4-digit hexadecimal constant F0A8. The result of the comparison is also available at the first three variables of the external output channel 7.

Shift register instruction
This instruction is identified by characters SFT and acts as a shift register of one or several 16-bit channels.

The SFT instruction has three associated input variables (each one being a binary variable or a combination of binary variables). The first to be programmed is the data input, the second is the clock input and the third the reset input. Finally, the SFT instruction is programmed in association with the start and end channels. The programming of these channels should be in ascending order, that is first the lower-order channel is programmed and then the higher-order one, except if the register is of 16 bits (one channel only) in which case the start and the end channels are the same. In their turn, the two channels must be contained in the same group of variables

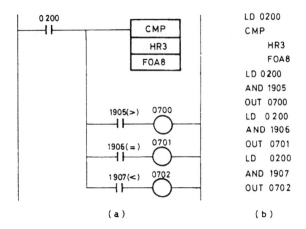

FIGURE 7.25 *Example of comparison instruction application.*

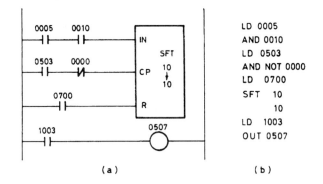

FIGURE 7.26 *Example of shift register instruction application.*

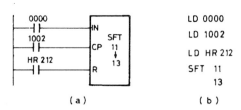

FIGURE 7.27 *Example of shift register instruction application.*

(internal output, internal output hold-up or external output). If the channel or channels belong to the internal output hold-up group, they keep their value when a power failure occurs. If the register variables are internal output variables, some or all of them can be transferred by the program to act upon external variables. Each bit is shifted inside the register, at the rising edge of the clock pulses.

Figure 7.26 represents a shift register of the 16 bits of channel 10 (ranging from 1000 to 1015), using binary variable 1003 as an external output. Figure 7.27 represents a shift register of 48 bits (ranging from 1100 to 1115, 1200 to 1215 and 1300 to 1315).

CHAPTER 8

Programming units and peripherals

8.1 Introduction

In chapter 7 we mentioned that one of the functions incorporated in microprocessor-based PLCs is the communication unit which allows the PLC to be connected to external electronic systems. The functions of these external electronic systems can be divided into two major classes:

1. **Applications development aids**
   - Program implementation in a given language.
   - Transferring the program to the active memory (RAM) of the PLC.
   - Programming passive memories (EPROM, EEPROM or FLASH) to be installed in the PLC.
   - Checking the program evolution during execution.
   - Program changes.
   - Fault location.
   - Program storage.

   These systems can be classified as programming units.

2. **PLC connections**
   - Remote systems to enhance the input and/or output capacity.
   - Remote peripherals for result display, parameter changes, etc.
   - Other PLCs to implement distributed control systems.
   - Computers.

8.2 Programming units

The solution of an industrial control problem using a PLC consists in translating operation specifications into a machine code program to be stored in memory. As seen in chapter 5, there are several specification languages to make the task of program design easier.

PROGRAMMING UNITS AND PERIPHERALS 307

Translation from these specification languages to the machine language is a complex task which is done by electronic devices, called programming units, which make use of computer aided design (CAD) techniques.

A programming unit can be defined, then, as an electronic system orientated towards program design, the transference of the programs to the PLC and the simulation of their behaviour. The many and various programming units available show different performance capabilities. These capabilities are discussed below.

Information display capability

As seen in chapter 5, the different specification languages of a PLC have very different requirements where the display of information is concerned. The instruction list language uses only alphanumerical characters, and if only one instruction at a time needs to be seen then only an alphanumerical

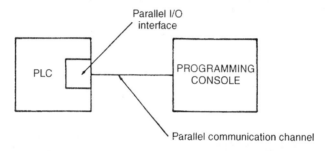

FIGURE 8.1 *Programming unit using parallel communication.*

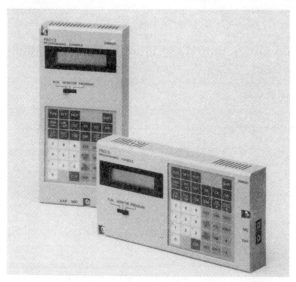

FIGURE 8.2 *Programming unit directly connectable to the PLC (courtesy of Omron).*

display is needed. On the other hand, the ladder diagram and function diagram languages need to represent certain structures which require a graphical (high resolution) display.

In the case of the instruction list language, the programming unit is a microcomputer with a display and a specialized keyboard which is available in two principal versions:

1. A unit which can be connected directly to the PLC, therefore requiring no cables for a parallel connection (Figure 8.1). An example is shown in Figure 8.2.
2. A unit that can be connected to the PLC by a cable in a serial connection (Figure 8.3). An example is shown in Figure 8.4.

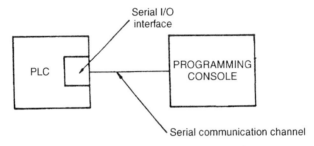

FIGURE 8.3 *Programming unit using serial communication.*

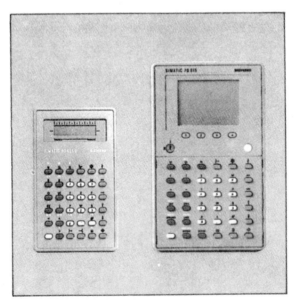

FIGURE 8.4 *Programming unit connectable to the PLC through a cable (courtesy of Siemens).*

FIGURE 8.5 *Computer-based programming unit (courtesy of Siemens).*

Programming units with graphical capabilities are really computers with hardware and software resources which make the implementation of complex programs in any language easier. Some PLC manufacturers supply their own computers as programming units (Figures 8.5 and 8.6) while others provide software for compatible personal computers.

FIGURE 8.6 *Computer-based programming unit (courtesy of Omron)*.

There are several advantages in using a compatible personal computer as a programming unit, namely.

1. For editing the program, including comments, to provide better documentation.
2. For disk storage.
3. For printing out distinct parameters such as the input variables, output and internal state variables and the instructions involving them.

8.3 Peripheral units

The increasing complexity of industrial sites and their production processes requires the capability to connect a PLC to electronic systems which increase the PLC's capacity and enlarge its range of application. The various communication problems associated with the implementation of such a connection are gradually being standardized by several international organizations. Below, some of the solutions with wider applicability are analyzed.

Figure 8.7 represents the connection of a data terminal using a standard serial connection (RS-232C, RS-422, etc.). The PLC must have a serial I/O interface unit and the appropriate subroutines in its monitor program.

PROGRAMMING UNITS AND PERIPHERALS 311

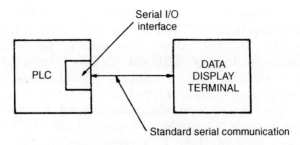

FIGURE 8.7 *Connection of a data terminal to a PLC using a standard serial communication.*

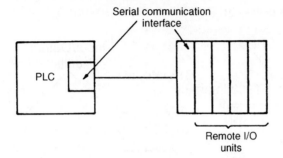

FIGURE 8.8 *Connection of a remote I/O unit to a PLC using a serial communication.*

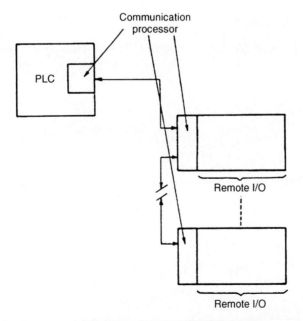

FIGURE 8.9 *Connection of several remote I/O units to a PLC using a serial communication.*

Figure 8.8 represents another application of great practical interest which consists in connecting remote I/O units. For that purpose, the PLC and the I/O unit have serial communication units, each of which coordinates the dialog either with its host or with the element connected to it. Given the presence of two communication units, the PLC acts upon the remote I/O unit with the same instructions as if it were directly connected to its buses.

Figure 8.8 is a particular case of a more general structure as shown in Figure 8.9 where the PLC is connected to several remote I/O units. In this situation, in each element there is a communication processor with the required software and hardware. The system is in fact a local area network (LAN) which can be structured in several different ways: periodic sampling, ring, bus, etc. A practical example is shown in Figure 8.10.

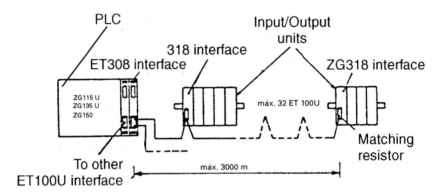

FIGURE 8.10 *Network of remote I/O units (courtesy of Siemens).*

APPENDIX 1
Standard logic symbols

A1.1 Introduction

The development of microelectronics during the 1970s made possible the fabrication of large and very large integrated circuits (LSI and VLSI) containing complex digital functional blocks. That development gave rise to a need for symbols to represent those blocks. Such symbols were put forward every time that a new functional block appeared, but the process lacked a set of rules to simplify the representation of new blocks when required.

By the end of the 1970s several institutions had begun to develop a set of standards for the representation of logic functions. This led to the creation of standard logic symbols adopted by the International Electrotechnical Commission (IEC) which is part of the ISO (International Standards Organization).

The general adoption of these symbols, in particular by integrated circuits manufacturers in their data sheets, facilitates the interpretation of digital systems diagrams and helps to develop the computer assisted design (CAD) of digital electronic systems. The electronics engineer should make an effort to learn the new symbolic language and to use it in circuit diagrams.

In subsequent sections the new standard symbols are analyzed and compared with their predecessors.

A1.2 Standard graphic symbol

The standard symbol of digital functional blocks adopted by the IEC is based on a vertical rectangle endowed with the appropriate symbols to show the function it executes.

Figure A1.1 shows the general standard symbol composition. At the top it shows the general qualifying symbol, indicating the function it performs. On

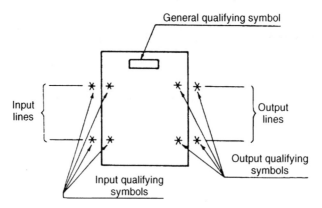

FIGURE A1.1 *Composition of the standard symbol of a binary logic element.*

the left we place the input lines and on the right the output lines, so that the direction of information flow is from left to right.

Each input pin can have one or two qualifying symbols specifying the way it acts on the circuit. Similarly, each output pin may have qualifying symbols specifying its characteristics. In the following sections, the different symbols associated with input and output pins are studied.

When a complex functional block is composed of several identical blocks, it can be represented by adjoining symbols and placing the general qualifying symbol only in the top one. Figure A1.2 shows four identical digital functions with two input variables and one output variable implemented in a single integrated circuit.

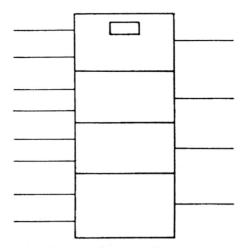

FIGURE A1.2 *Standard symbol of a functional block with four identical logic elements.*

Complex functional blocks often comprise several identical circuits with some inputs specific to each and some inputs common to all. In these cases (a characteristic example is the multiple multiplexer block), a common block, such as the one shown at the top of Figure A1.3a, is added to the adjoining rectangular symbols. This common block is connected to the inputs common to all blocks. Figure A1.3a shows the symbol of a functional block with four identical digital functions, each with two unique inputs and one common input. Figure A1.3b shows the circuit whose equivalent symbol is that of Figure A1.3a.

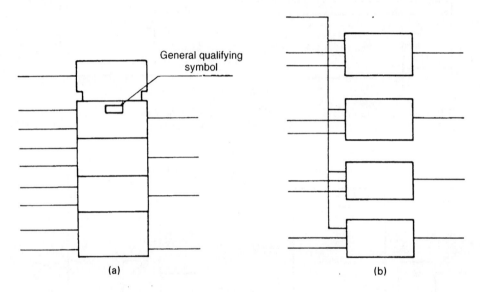

FIGURE A1.3 (a) *Standard symbol of a functional block with four identical digital functions that have one common input variable.* (b) *Digital circuit equivalent to the symbol shown in* (a).

Some functional blocks comprise several independent digital functions and a function which is dependent on the output of the others, as in Figure A1.4a. The representation can be simplified as indicated in Figure A1.4b.

Given that variable a is common to the three functions, it is connected to the top block. To indicate that output $f3$ of the bottom block depends on variable h and on the outputs $f1$ and $f2$ of the other blocks, we separate it from the others by two parallel lines and only variable h is indicated.

Sometimes, some inputs of a functional block are combined by means of a certain logic function different from that executed by the block. A possible representation is given in Figure A1.5a where the logic product of variables c and d is applied to the second block. Figure A1.5b indicates a simplified representation with embedded symbols.

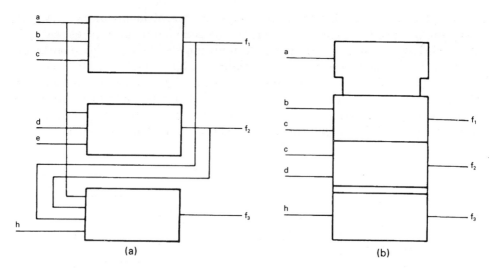

FIGURE A1.4 (a) *Digital circuit with two independent functions and another function which combines them.* (b) *Standard symbol equivalent to the circuit shown in (a).*

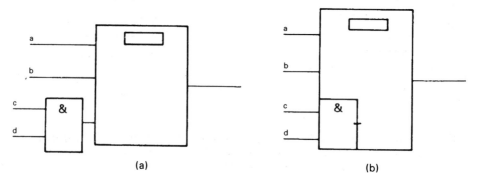

FIGURE A1.5 (a) *Interconnected standard logic symbols.* (b) *Equivalent embedded symbols.*

A1.3 Symbols associated with inputs and outputs

In the previous section, the general qualifying symbol used to represent different logic elements is analyzed. A set of standard symbols should provide an adequate description of the different types of pin and how signals present on those pins act on the circuit.

The symbols associated with the pins of the logic elements may refer to the logic function they execute or to their electrical characteristics. The logic function of a pin may be specific to a certain type of functional block or identical for all the blocks.

Specific functions are studied in subsequent sections and those of general use are analyzed below.

Figure A1.6 shows the logic negation symbol associated with input and output pins. This symbol consists of a circle and, for an input (Figure A1.6a), the circle indicates that an external 0 is converted into a 1 applied to the block, and vice versa for an external 1. Similarly, a circle at the output of a functional block (Figure A1.6b) indicates that the function executed by the block is negated before being connected to the output.

FIGURE A1.6 *Standard inversion symbol for inputs* (a) *and outputs* (b).

The circle used in Figure A1.6 to represent the logic negation does not indicate if the underlying logic is positive nor negative. Therefore, a triangle has been adopted to represent negation in positive logic, as shown in Figure A1.7.

FIGURE A1.7 *Standard inversion symbol for inputs* (a) *and outputs* (b) *using positive logic.*

In sequential systems there are inputs acting by level (as, for example, the output enable or the selection of operating mode) and others acting on edges or transitions (for example, a clock or control input). The former use the symbols of Figure A1.6 to represent negation, and simply do not have a symbol when there is no negation.

To distinguish edge-sensitive (or dynamic) inputs from level-sensitive (static) inputs we add to the former an internal symbol consisting of an isosceles triangle. Figure A1.8a shows an input responding to 0-to-1 transitions and Figure A1.8b another responding to 1-to-0 transitions. Under the assumption of positive logic, the circle can be replaced by a triangle as indicated in Figure A1.9.

FIGURE A1.8 *Dynamic inputs.* (a) *Positive- or rising-edge-sensitive input.* (b) *Negative- or falling-edge-sensitive input.*

FIGURE A1.9 *Dynamic inputs using positive logic.*

Some inputs have hysteresis, that is the 0-to-1 and 1-to-0 output transitions are executed at different input voltage levels. These inputs are indicated by the internal symbol shown in Figure A1.10.

It is also worth looking at the electrical characteristics of outputs. The internal symbols shown in Figures A1.11 and A1.12 are used to represent these characteristics. Figure A1.11 shows the symbol of a tri-state output

FIGURE A1.10 *Hysteresis input.*

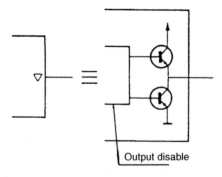

FIGURE A1.11 *Tri-state output standard symbol.*

implemented with bipolar transistors as indicated on the right, and Figure A1.12 shows the open-circuit output which in turn has two variants. Figure A1.13 shows an open-collector output (L-type open-circuit) (open drain in the case of MOS transistors) and Figure A1.14 shows an open-emitter output (H-type open-circuit) (open source in the case of MOS transistors). In addition, the open-circuit outputs may incorporate an internal resistor as a passive load for the transistor. Figure A1.15 shows a passive pull-up output and Figure A1.16 a passive pull-down output.

FIGURE A1.12 *Open-circuit output standard symbol.*

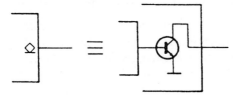

FIGURE A1.13 *Open-collector or open-drain output standard symbol.*

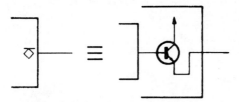

FIGURE A1.14 *Open-emitter or open-source output standard symbol.*

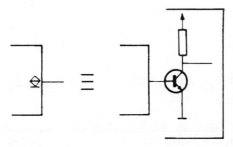

FIGURE A1.15 *Passive pull-up output standard symbol.*

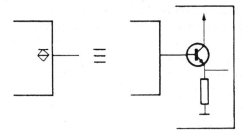

FIGURE A1.16 *Passive pull-down output standard symbol.*

A1.4 Standard representation of combinational systems

A1.4.1 Generalities

Combinational systems may be classified into two major categories:

1. Those where there is no dependency amongst inputs, i.e. the state of each input does not determine the way the circuit acts upon the other inputs. These systems execute certain logic functions and are called logic gates.
2. Those where there is dependency between some inputs. Typical examples are selection inputs and enable inputs. The implementation of these systems gives rise to complex functional blocks manufactured as medium-scale integrated circuits (MSI) or as subsystems of more complex systems implemented as large or very large scale integrated circuits (LSI and VLSI).

Both types of combinational system are analyzed below.

A1.4.2 Logic gates

The logic symbol used is the general rectangle, with the qualifying symbol of the implemented function placed at the top. Table A1.1 shows the qualifying symbols of the most important logic functions. Figures A1.17 and A1.18 are the diagrams of two combinational circuits implemented with NAND and NOR gates using the non-standard symbols (Figures A1.17a and A1.18a) and the standard symbols (Figures A1.17b and A1.18b).

A1.4.3 Qualifying symbols for complex combinational functional blocks

Combinational functional blocks are complex combinational circuits in which, in general, some inputs modify the way that the circuit operates upon other inputs. The modifying inputs are known as control inputs and the modified as

APPENDIX 1

TABLE A1.1 *General qualifying symbols for logic functions*

Symbol	Logic function
&	AND function or AND gate
≥ 1	OR function or OR gate
= 1	Exclusive-OR gate of two inputs: its output is a one if and only if one of its inputs is a one.
=	Logic identity function. The output is a 1 when all the inputs stand at the same state.
= m	The output is a one if and only if m inputs are one.
> $n/2$	The output is a one if the majority (more than half) of the inputs are one.
≥ m	Threshold function: its output is a one if a minimum of m inputs are one.
2k	The output is a one whenever there is an even number of inputs equal to one.
2k + 1	As above for an odd number of inputs.
▷ or ◁	Buffer: element with greater than usual output driving capability. The symbol indicates the direction of the signal flow.

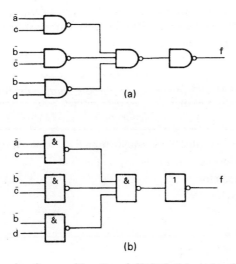

FIGURE A1.17 *Example of a combinational digital circuit implemented with NAND gates.* (a) *Using non-standard symbols.* (b) *Using standard symbols.*

information or operating inputs. We then say that there is a dependency relationship between the two types of input. Although there are various types of dependency relationship which differ in the way the modification is done, there is a general rule for their representation which is: 'The control input is indicated by a letter, corresponding to the type of dependency relationship, followed by a number. Each affected input has the same number'.

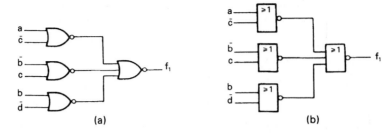

FIGURE A1.18 *Example of a combinational digital circuit implemented with NOR gates.* (a) *Using non-standard symbols.* (b) *Using standard symbols.*

Similarly, in complex combinational circuits there are inputs which modify the result or the output variables of the circuit and the same rule applies. This rule will be easily understood by means of several examples in the following sections.

A general qualifying symbol is assigned to each type of functional block to define the function it executes. Table A1.2 shows the symbols for the most important combinational functional blocks, including passive or read-only memories (ROM, PROM, and EPROM), programmable logic arrays (PLAs) and programmable array logic devices (PALs).

TABLE A1.2 *General qualifying symbols for combinational functional blocks*

Symbol	Logic function
X/Y	Code converter (for example, from BCD to seven segments)
HIPRI/BIN	Priority encoder with binary output
COMP	Comparator
MUX	Multiplexer
DEMUX or DX	Demultiplexer
Σ	Adder
P − Q	Subtractor
CPG	Look-ahead carry generator
ALU	Arithmetic logic
Π	Multiplier
ROM	Fixed read-only memory
PROM	Programmable read-only memory
EPROM	Erasable programmable read-only memory
PLA	Programmable logic array
PAL	Programmable array logic

APPENDIX 1

Major dependency relationship types and ways of combining them are described below.

AND dependency relationship [G(AND)]

This relationship shows a logic product between the input indicated by the letter G followed by a number a and the information inputs, or outputs, indicated by a.

Figure A1.19a shows an AND dependency relationship between input n and the output of a functional block by the inclusion of an AND gate. Figure A1.19b indicates the standard symbol, assigning to the input n the symbol Ga and to the output the symbol a. It must be emphasized that Figure A1.19 is only a graphical way of representing AND dependency. In practice, function AND does not need a separate gate since it can be performed in conjunction with the other functions of the block.

A typical example of AND dependency between inputs and outputs are demultiplexers. Figure A1.20 shows a demultiplexer with eight output

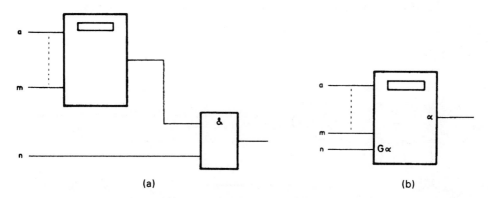

FIGURE A1.19 *AND dependency relationship between one input and one output.*

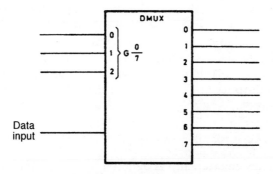

FIGURE A1.20 *Standard symbol of a demultiplexer with eight output channels.*

channels. We assign symbols G0/7 to inputs 0, 1 and 2. These symbols indicate that each input combination corresponding to numbers from 0 to 7 affects the corresponding output in accordance with the function AND(G).

Figure A1.21 shows the AND dependency relationship between two inputs, which is similar to the above-mentioned relationship between an input and an output. A typical example of this relationship are multiplexers. Figure A1.22 shows a 4-channel multiplexer where each combination of the selection variables 0 and 1 selects one of the four input channels. To indicate this, we assign symbol G0/3 to these variables and the corresponding numbers 0, 1, 2 and 3 to each input channel.

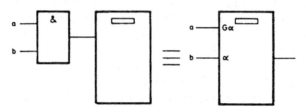

FIGURE A1.21 *AND dependency relationship between two inputs.*

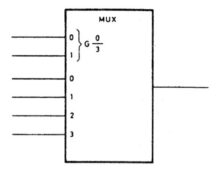

FIGURE A1.22 *Standard symbol of a four-input channel multiplexer.*

Summarizing, we can say that the AND dependency relationship is as follows. When a $G\alpha$ input stands at its 0 internal state (the state inside the block without taking into account a possible external negation), all inputs and outputs indicated by α stand at 0. If, on the other hand, $G\alpha$ is a 1, all the affected inputs stand at the logic level fixed by the corresponding input pins and the affected outputs at the logic level determined by the inputs and the block logic function.

OR dependency relationship [V(OR)]

The OR dependency relationship is denoted by a V. When an input $V\alpha$ is a 1, all the internal inputs and outputs indicated by α are set to 1. If, on the other hand, Vα is a 0, all inputs and outputs affected by Vα stand at their normally defined states.

Negation dependency relationship [N(NEGATE)]

The negation dependency relationship represents an exclusive OR function between the inputs indicated by letter N followed by a numerical symbol α and the inputs or the outputs indicated by the same numerical symbol α. Figure A1.23 shows this exclusive OR function between an input and an output and its indication by means of the dependency relationship N.

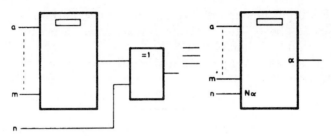

FIGURE A1.23 *Negation dependency relationship between one input and one output.*

Figure A1.24 shows an example of the dependency relationship N using the 4-channel multiplexer of Figure A1.22 to which an input has been added to act upon the output by means of the relationship N. If input N4 is a 0, output 4 takes the value determined by the state of the remaining inputs. If N4 is a 1, output 4 takes the complement of the value determined by the state of the remaining inputs. Therefore, N4 is a control variable of the multiplexer output negation.

The reader can easily obtain the definition of the N relationship.

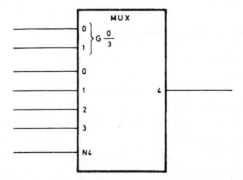

FIGURE A1.24 *Standard symbol of a 4-channel multiplexer with a control variable of the negation or inversion of the output.*

Enable/disable dependency relationship [EN(ENABLE)]

As the name suggests, EN_a affects the inputs labelled with a by disabling their action when EN_a stands at 0 and enabling their action when it stands at 1. If EN_a affects outputs a, it can perform one of the three following functions:

1. If the outputs are of the open-collector type, they are turned off.
2. If the outputs are tri-state outputs, they are set at the high impedance state.
3. If they are normal outputs they are set at the internal 0 state. In this case the EN dependency relationship is identical to G.

When an EN input has no number a, it affects all outputs of the block in the way indicated above. This function is much used, especially in combinational or sequential functional blocks.

Figure A1.25 shows a 4-input channel multiplexer with a tri-state output and an enable input EN.

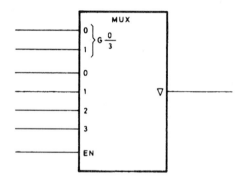

FIGURE A1.25 *Symbol of a 4-channel multiplexer with an enable input of the tri-state output.*

Operating mode relationship [M(MODE)]

M-type inputs may affect either information inputs or outputs. This type of relationship is particularly useful in sequential functional blocks (although it also has application to combinational functional blocks) and consequently we refer the reader to section A1.5.3.

Relationship M affects the outputs of combinational systems by selecting their operating mode. Figure A1.26 shows an example of four drivers, with a common enable input EN and an operating mode input M1. Input M1 selects the operating mode of the output in such a way that if it is a 1 the output is a tri-state output and if it is a 0, it is an open-collector output. This is indicated by labelling the output with the symbols $1\triangledown/\bar{1}\diamondsuit$.

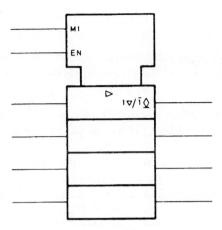

FIGURE A1.26 *Symbol of four drivers with operating mode and enable control inputs.*

Interconnection dependency relationship (Z)

The interconnection dependency relationship indicates that there is a connection between terminals labelled $Z\alpha$ and α. It is used to simplify the representation of complex functional blocks.

Figure A1.27 shows an example: a combinational system that not only executes a certain function with its input variables but also executes the AND function of four of them. In order to avoid the explicit representation of the connections, the inputs are labelled Z1 to Z4 and the AND gate inputs are labelled 1 to 4.

Address dependency relationship (A)

The address dependency relationship is used to represent the selection of a logic element in a multidimensional array. The most frequent case is the

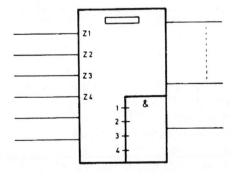

FIGURE A1.27 *Example of interconnection dependency relationship.*

random access memory which has a given number of locations with a given number of bits each. The selection can be done in two ways:

1. By means of an independent terminal for each location.
2. By means of a number n of terminals, such that 2^n is the total number of locations.

The first is used when the memory has a small number of locations (in general no more than eight). The second method is more frequently encountered.

Figure A1.28 shows a programmable read-only random access memory (PROM) of four locations of eight bits each with a tri-state output (controlled via the enable input, EN) and four address terminals A1 to A4.

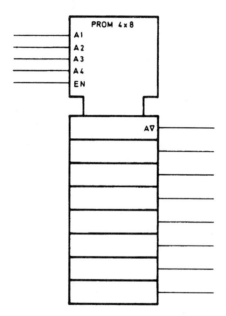

FIGURE A1.28 *Symbol of a programmable read-only memory (PROM) with four locations of eight bits and one address pin for each.*

If the memory is enable (EN = 1), its output is the data of the location corresponding to the terminal Ai which has a logic 1 (only one A input may have a logic 1 at any time). Letter A located at the output indicates that its logic value is selected by the Ai inputs.

Figure A1.29 shows a memory with the same number of locations and bits per location as the one represented in Figure A1.28. The only difference is that the address selection is done by two bits ($2^2 = 4$) in such a way that the accessed location each time has an address equal to the combination present in such bits.

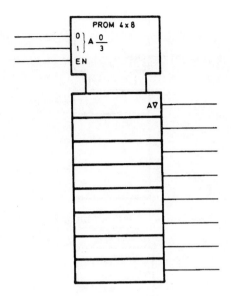

FIGURE A1.29 *Symbol of a programmable read-only memory (PROM) with four locations of eight bits and two address pins.*

Combination of dependency relationships

Dependency relationships can be combined to affect the operation of a functional block. In practice, the combination can be achieved in two ways:

1. **Simultaneity combination**. This is specified by the appropriate symbols separated by commas. We use it when an input acting in just one way on the contents of a functional block is affected by the logic state of several inputs simultaneously or when the logic state of an output depends on several inputs simultaneously. Figure A1.30 shows a combinational functional block with a tri-state control EN2 and an enable G1 input affecting simultaneously the output, indicated by symbols 1, 2▽. Digit 2 next to the symbol ▽ indicates that when EN2 is a logic 0, the output stands at the third state, and when it is a 1, the output is set at the corresponding state depending on the value of the information inputs and on the function of the block. Number 1 separated from number 2 by a comma indicates that when EN2 is a 1, the output state is a logic 0 if G1 is in such state.

2. **Exclusion combination**. This is specified by the appropriate symbols separated by a stroke. It appears in combinational systems when they have operating mode inputs. A typical example is shown in Figure A1.26. The output is indicated as 1▽/$\bar{1}$◇ because it is a tri-state output (1▽) if M1 is a 1 and an open-collector output ($\bar{1}$◇) if M1 is a 0.

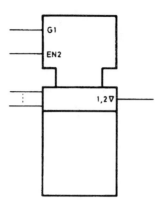

FIGURE A1.30 *Simultaneity combination of dependency relationships.*

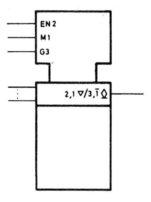

FIGURE A1.31 *Simultaneity and exclusion combinations of dependency relationships.*

Both simultaneity and exclusion combinations can be combined too. Figure A1.31 shows a functional block similar to that of Figure A1.26 to which an AND dependency relationship input (G3) has been added. G3 is enabled when M1 is a 0, while EN2 is enabled when M1 is a 1.

A1.5 Sequential system standard representation

A1.5.1 Introduction

Sequential systems may be divided into two major categories:

1. Basic memory elements which have two internal states and constitute synchronous and asynchronous flip-flops. In spite of their simplicity, flip-flops have dependency relationships among their inputs.

APPENDIX 1   331

2. Complex sequential systems which have a large number of operating modes. These are functional blocks that constitute medium-scale integrated circuits (MSI) or are subsystems of large or very large scale integrated circuits (LSI or VLSI).

Just as in the case of combinational systems, there are standard symbols and dependency relationships for sequential systems. These are studied below.

A1.5.2 Sequential system symbols

Flip-flops have no special symbols because they can be identified by their inputs and outputs. Complex sequential functional blocks have specific qualifying symbols. Table A1.3 shows those whose label has already been standardized.

TABLE A1.3 *General qualifying symbols for sequential functional blocks*

Symbol	Function
SGRm	m bit shift register
CTRm	m bit binary counter; cycle length = 2^m
CTRDIVm	Counter with cycle length = m
RAM m1xm2	Random access memory with $m1$ positions of $m2$ bits
CAM m1xm2	Content addressable or associative memory with $m1$ positions of $m2$ bits
FIFO m1xm2	First in-first out memory with $m1$ positions of $m2$ bits
⎍	Retriggerable monostable
1⎍	Non-retriggerable monostable
⎍G⎍	Pulse generator, astable or clock

Besides the general qualifying symbols, sequential systems have specific symbols for inputs and outputs. Table A1.4 indicates the most frequently used. This table should be referred to frequently when dependency relationships are studied in the next section.

In addition to the symbols of Table A1.4, those described in section A1.3 are also used in sequential systems.

A1.5.3 Sequential system dependency relationships

The dependency relationships studied in section A1.4.3 for combinational systems also apply to sequential systems, namely the AND relationship [G(AND)], the OR relationship [V(OR)], the inversion relationship [N(NEGATE)], the enable/disable relationship [EN(ENABLE)], the

TABLE A1.4 *Sequential system input and output symbols*

Input or output symbol	Function
J and K	Inputs to a J–K flip-flop
D	Input to a D flip-flop and, in general, data input to any sequential functional block
C	Control or clock input to a sequential system
S	Set input to a flip-flop or to a sequential functional block
R	Reset input to a flip-flop or to a sequential functional block
→	Shift right pulse input
←	Shift left pulse input
+	Counting up pulse input
−	Counting down pulse input
CT = m	Content input. When active, causes the content of the memory elements of a sequential functional block to take on the binary number equivalent to the decimal number m.
CT = m	Content output. When stands at level 1 it indicates that the content of the memory elements of a sequential functional block is equal to the binary number equivalent to the decimal m.

operating mode relationship [M(MODE)], the interconnection relationship (Z) and the address relationship [A(ADDRESS)].

In sequential systems, however, there are specific dependency relationships such as reset (R), set (S) and control (C). Furthermore, the operating mode relationship (M) has different aspects when used with sequential systems. These specific relationships are described in the following sections.

Reset (R) and Set (S) dependency relationships

These relationships define the action of certain inputs of sequential functional blocks which set the memory elements to a given state.

The reset input [R(Reset)] sets the memory elements to state 0 and the set input [S(Set)] sets them to state 1.

The most significant example of such a dependency relationship is the asynchronous R–S latch, whose logic symbol is shown in Figure A1.32.

Figure A1.33 shows the logic symbol of a synchronous functional block with the reset input (R) of all the memory elements on the common part.

Control dependency relationship (C)

This dependency relationship is specific to the memory elements or synchronous flip-flops. The inputs denoted by C control the action of the other

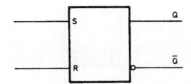

FIGURE A1.32 *Standard logic symbol of an R–S flip-flop.*

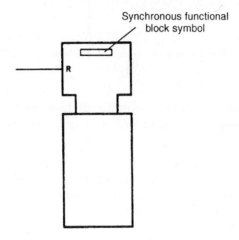

FIGURE A1.33 *Standard logic symbol of a synchronous functional block with a common reset (R) input variable.*

inputs, known as information inputs, on the state of the flip-flop. In non-standard symbols, this input is labelled T.

There are two types of C input depending on the way they execute the control:

1. **Control by level input**. When this input is at a certain level it allows the information inputs to affect the state of the flip-flop. It is labelled with the letter C followed by a decimal number.

    Figure A1.34 shows a level-triggered D flip-flop. The synchronization input has the label C followed by the number α and the information input D has the same number α.

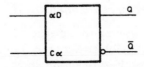

FIGURE A1.34 *Standard symbol of a level-triggered D flip-flop.*

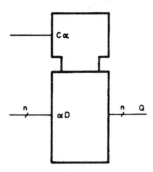

FIGURE A1.35 *Standard symbol of a parallel I/O register using level-triggered D flip-flops.*

Figure A1.35 shows another example of how the control relationship C can be used. It shows a register of n D level-triggered flip-flops with a single pulse input.

2. **Edge control or dynamic input**. When a change of level, which can be from 0 to 1 or from 1 to 0, takes place at this input, the information inputs act on the state of the flip-flop. These inputs are also labelled with C followed by a number and preceded by the edge indicator or dynamic symbol, represented in Figure A1.8.

Figure A1.36 shows a positive- or rising-edge-triggered D flip-flop and Figure A1.37 shows a falling- or negative-edge-triggered register with n flip-flops.

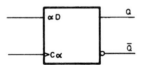

FIGURE A1.36 *Standard symbol of an edge-triggered D flip-flop.*

In certain functional blocks, e.g. counters, label C is not necessary because the function of the corresponding pin is indicated by the dynamic input symbol (see Figures A1.38 to A1.44).

Operating mode dependency relationship (M)

Just as in combinational systems, the operating mode relationship M defines the way that certain input variables act on a functional block or the way they control the output variables. Figure A1.38 shows, as an example of how the operating mode acts on an input variable, a counter with a mode input Mα.

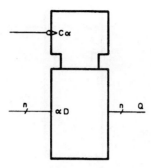

FIGURE A1.37 *Standard symbol of a parallel I/O register using edge-triggered D flip-flops.*

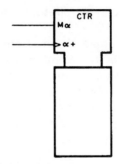

FIGURE A1.38 *Standard symbol of an up synchronous counter with an operating mode (M) input variable.*

When $M\alpha$ is at logic 1 the counter is allowed to count up the pulses applied to input α.

Figure A1.39 shows an operating mode example affecting an output variable of a counter. When the input $M\alpha$ is at logic 1, the output $\alpha Ct = 9$ is set to 1 if the content of the counter is 1001 (equivalent to decimal number 9).

Combination of dependency relationships

The dependency relationships can also be combined in sequential functional blocks. The greater complexity of sequential, as compared with combinational, functional blocks means that such combinations appear frequently.

1. **Simultaneity combination**. This is specified by the appropriate symbols separated by commas. It appears in the following situations:
    - where an input affecting in only one way the content of a functional block depends on the state of several inputs simultaneously;

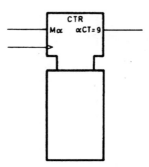

FIGURE A1.39 *Standard symbol of a counter with an operating mode input variable and a content indication output.*

- where the state of an output depends on various inputs simultaneously.

Figure A1.40 shows an example of simultaneity combination in sequential systems. The lower input of the counter is a dynamic input (section A1.3) and the + sign specifies that it counts the pulses up. In this functional block, counting only takes place if the mode input $M\alpha$ and the inhibition input $G\beta$ are both at logic 1. To specify this behaviour, the input to which the counting pulses are applied is labelled as $\alpha, \beta+$.

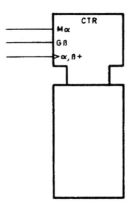

FIGURE A1.40 *Standard symbol of an up-counter with an operating mode (M) and an enable (G) input variables.*

2. **Exclusion combination**. This is also specified by the appropriate symbols separated by strokes. It appears in the following situations:
    - where an input affecting a functional block in several ways is conditioned by the logic value of one or more inputs in different ways;
    - where an output may assume different logic states depending on the logic value of one or more inputs.

Figure A1.41 shows an example of the exclusion combination. The lower input of the counter has the symbol of a dynamic input followed by symbols $\alpha+/\bar{\alpha}-$. This means that if input $M\alpha = 1$, the pulses are counted up, and if input $M\alpha = 0$ the pulses are counted down.

In sequential systems, both ways of combining the dependency symbols are often found together. Figure A1.42 shows a simple example which combines the symbols studied in Figures A1.40 and A1.41. The counter of Figure A1.42 counts the rising edges applied to the dynamic input in the following ways:

- Up if $M\alpha = 1$ and $G\beta = 1$.
- Down if $M\alpha = 0$ and $G\beta = 1$.

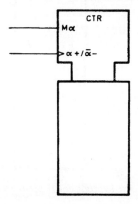

FIGURE A1.41 *Standard symbol of an up/down counter with a counting direction operating mode input variable (M).*

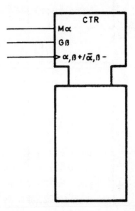

FIGURE A1.42 *Standard symbol of an up/down counter with a counting direction operating mode (M) and an enable (G) input variables.*

In the simultaneity and exclusion combinations, factoring can be applied in order to simplify the standard symbols. As an example, by factoring Figure A1.42 we obtain the simplified version, Figure A1.43.

The simultaneity and exclusion combinations can also be applied to the output variables. Figure A1.44 shows, as an example, a counter with one operating mode input Mα and a content indication output with the symbols $\bar{\alpha}$CT = 0/αCT = 15. These symbols specify that the content indication output is at a logic 1 if either of the following situations arises:

1. Input Mα = 0 and the content of the counter is equivalent to decimal number 0.

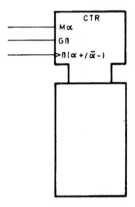

FIGURE A1.43 *Standard symbol equivalent to that of Figure A1.42.*

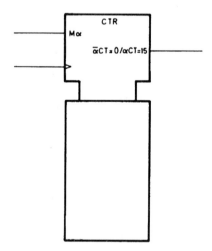

FIGURE A1.44 *Standard symbol of a synchronous counter with an operating mode input variable which selects the function of the content indication output.*

2. Input $Ma = 1$ and the content of the counter is equivalent to decimal number 15.

A1.5.4 Practical examples of sequential systems

In this section, different sequential functional blocks are analyzed to prove that the representation rules indicated above avoid the use of logic tables or operation tables and lead to self-explanatory diagrams.

The different sequential functional blocks have the common characteristic of including several flip-flops, sometimes together with a combinational circuit. Both are characterized by having specific and common inputs. The general symbol is shown in Figure A1.45.

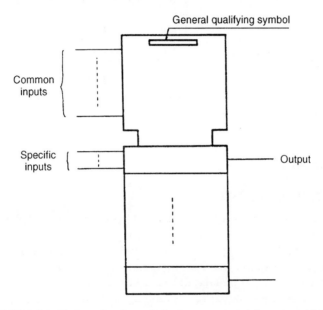

FIGURE A1.45 *Standard symbol of a sequential functional block.*

Parallel registers

Parallel registers are the simplest synchronous functional blocks. They consist of n flip-flops with a common set of control lines and all the information inputs and outputs available in parallel.

Figure A1.46 shows the general symbol of a parallel register, characterized by:

1. The absence of a general qualifying symbol.
2. Having n information inputs and n information outputs.

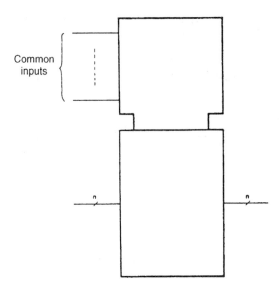

FIGURE A1.46 *Standard symbol of a parallel I/O register.*

3. The presence of common control inputs, among which the most usual are:
   - level- or edge-triggered data input control (C);
   - data input enable (G);
   - output enable (EN);
   - asynchronous reset (R).

Figure A1.47 shows an 8-bit parallel register made of D flip-flops, with a data input enable pin (clock enable) (G1) and a data input control (C2). The data present at the inputs of the eight flip-flops are stored when G1 is a 1 and a rising edge is applied to C2 simultaneously. This behaviour is indicated by the set of symbols 1,2D located at the input terminals.

Figure A1.48 shows a parallel register of 4 bits, with an asynchronous reset terminal (R), a tri-state output enable pin (EN) and a dynamic control input (C1). The data present at the inputs are stored in the register when a rising edge is applied to input C1 and R is a 0. To indicate this behaviour, symbol 1D is placed at the parallel input of the first flip-flop.

Parallel registers are combined with different types of combinational functional block like decoders and multiplexers.

Figure A1.49a shows a circuit formed by a 4-bit parallel register whose inputs are connected to a quad multiplexer with two input channels and a select input variable. Depending on whether this variable is a 0 or a 1, the data from terminals A_1 to D_1 or A_2 to D_2, respectively, will appear at the register inputs.

APPENDIX 1

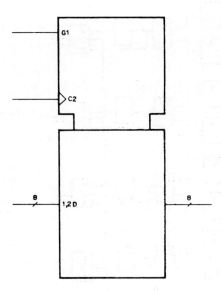

FIGURE A1.47 *Standard symbol of an 8-bit parallel I/O register with an information-entering enable input variable.*

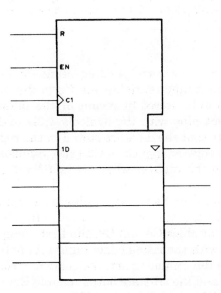

FIGURE A1.48 *Standard symbol of a 4-bit parallel I/O register with an asynchronous reset and a tri-state output control input variables.*

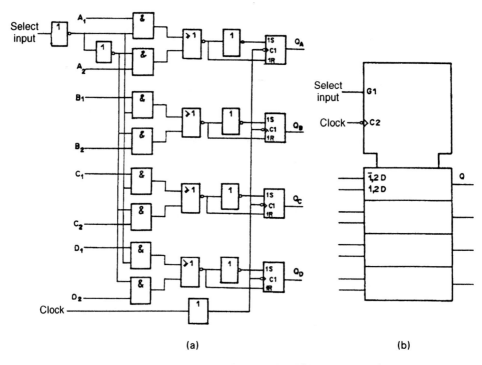

FIGURE A1.49 (a) *Logic circuit of a 4-bit parallel I/O register combined with four 2-channel multiplexers.* (b) *Corresponding standard symbol.*

Figure A1.49b shows the corresponding standard symbol. Terminal G1 corresponds to the select input variable and C2 to the dynamic control input which cause the data to be stored by means of the falling edges. Each block element has two input pins with the symbols $\bar{1},2D$ and $1,2D$, respectively. Symbols $\bar{1},2D$ indicate that the data are stored in the register when G1 is a 0 (i.e. $\bar{1}$) and an active edge is applied to C2 (2D). Symbols $1,2D$ indicate that the data are stored in the register when G1 is a 1 (1) and an active edge is applied to C2 (2D).

Figure A1.50a shows an 8-channel multiplexer with a register for the select variables and another for the input variables. The circuit has independent control variables C8 and C9, the first being active with the zero level and the second with the rising edge. Figure A1.50b shows the standard symbol which has, in the common part, the control variables C8 and C9, the selection variables G and the tri-state output enable EN.

The output enable EN consists of the AND logic function (&) of the three inputs $\overline{EN1}$, $\overline{EN2}$ and EN3.

The select variables S_0, S_1 and S_2 are connected to the inputs of a

APPENDIX 1

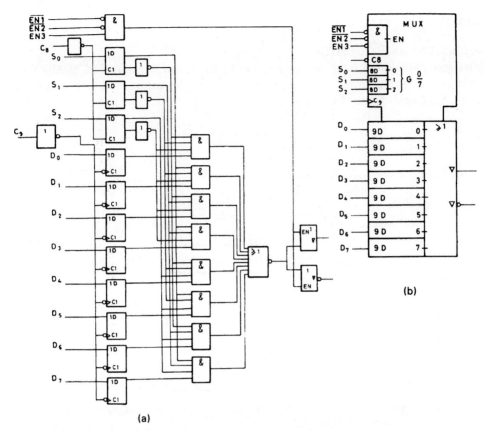

FIGURE A1.50 (a) *Logic circuit of an 8-channel multiplexer with registered select and data variable inputs.* (b) *Corresponding standard symbol.*

3-bit parallel register. They enter the register when input C8 is a 0 and are stored when this input is a 1. This action is indicated by label 8D placed at the select inputs and the inversion symbol before the C8 control input.

The data inputs D_0 to D_7 are connected to the inputs of a parallel register and are stored when a rising edge is applied to C9. Such action is indicated by the label 9D in the corresponding inputs. The outputs of this register are selected by the outputs of the selection register through an AND(G) dependency relationship.

To indicate this action we place labels G0/7 at the outputs of the select variable register and the numbers 0 to 7 at the outputs of the data register. For instance, if variables G are at levels $101_2 = 5_{10}$, the output of flip-flop 5 is selected to appear at the OR gate output.

Counters

ASYNCHRONOUS COUNTERS

The general symbol is shown in Figure A1.51. It is characterized by the fact that the counting input is not part of the common inputs but is the control input C of one of the flip-flops.

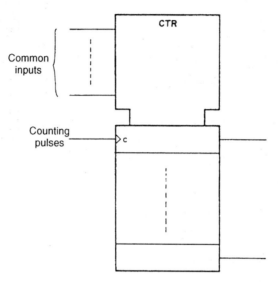

FIGURE A1.51 *General standard symbol of an asynchronous counter.*

Let us first consider how to obtain the logic symbol starting from the operating tables. An asynchronous decimal (BCD) counter with an asynchronous reset and counting disabling input is assumed. This counter counts the falling edges of the input signal. Table A1.5 shows the action of the different inputs and Table A1.6 indicates the counting code.

The standard symbol of Figure A1.52 has all the information of Tables A1.5 and A1.6 and renders those tables unnecessary. The general symbol CRTDIV10 indicates that it is a divide-by-10 counter, i.e. it counts from 0 to 9 (the absence of the counting code means that it is natural BCD). Placing

TABLE A1.5

R	G1	1C+	Q_3	Q_2	Q_1	Q_0
1	X	X	0	0	0	0
0	0	X	Counting disable			
0	1	↓	Count			

TABLE A1.6

Q_{3t}	Q_{2t}	Q_{1t}	Q_{0t}	Q_{3t+1}	Q_{2t+1}	Q_{1t+1}	Q_{0t+1}
0	0	0	0	0	0	0	1
0	0	0	1	0	0	1	0
0	0	1	0	0	0	1	1
0	0	1	1	0	1	0	0
0	1	0	0	0	1	0	1
0	1	0	1	0	1	1	0
0	1	1	0	0	1	1	1
0	1	1	1	1	0	0	0
1	0	0	0	1	0	0	1
1	0	0	1	0	0	0	0

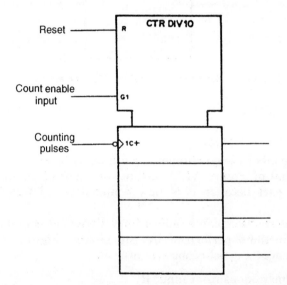

FIGURE A1.52 *Standard symbol of an asynchronous natural BCD counter with asynchronous reset (R) and counting enable (G1) control input variables.*

the counting input in the first flip-flop indicates that it is an asynchronous counter. Symbol 1C+ specifies that G1 has to be a 1 in order for counting to take place, and counting is upwards (+).

Figure A1.53 shows, as an example, the logic symbol of the 4-bit natural binary asynchronous counter integrated circuit (4-bit ripple clock) 7493. This circuit has a flip-flop constituting a divider-by-two and three more flip-flops forming a divider-by-eight. The divider-by-two has a falling-edge-triggered counting input, as does the divider-by-eight. The counter has two inputs which, when both at 1 (function &), reset (R) all the flip-flops.

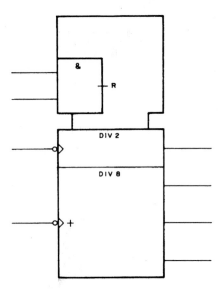

FIGURE A1.53 *Standard symbol of 7493 asynchronous counter.*

SYNCHRONOUS COUNTERS

The general symbol of synchronous counters is shown in Figure A1.54. It differs from that of Figure A1.51 in that the input of the counting pulses is in the upper part because it is now common to all the flip-flops of the counter.

Synchronous counters can handle a large variety of common inputs according to the functions performed by the circuit. Figure A1.55 shows the standard symbol of a synchronous counter with:

- an asynchronous reset input R;
- an operating mode input M1 to select up-counting (specified by symbols 1+ at the synchronization pulse input) or paralleling input (specified by symbol $\bar{1}$ at the input of each flip-flop);
- one enable input G2 which enables both parallel input and counting when it is 1;
- one clock or control input executing both the parallel input (specified by symbols C3) and the counting (specified by symbols 2, 1+ indicating that terminals G2 and M1 should be a 1 simultaneously). Since counting and parallel input are alternative actions, this input is labelled C3/2,1+;
- one parallel data input specified by symbols $\bar{1},2,3D$ which shows that in order to execute this action it is necessary to have, simultaneously, M1 at 0 ($\bar{1}$), G2 at 1 (2) and an active edge applied to C3 (3D).

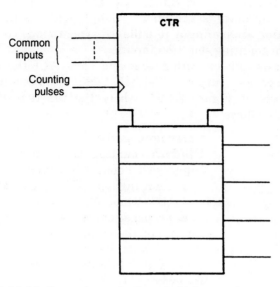

FIGURE A1.54 *General standard symbol of a synchronous counter.*

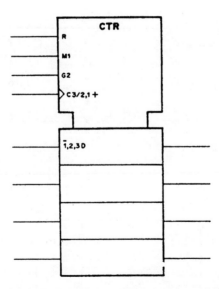

FIGURE A1.55 *Standard symbol of a synchronous counter with an asynchronous reset (R) input variable, a counting/parallel operating mode (M1) input variable, and a counting and parallel entering enable input variable (G2).*

A practical example is the integrated circuit 74191 constituting a 4-bit natural binary up/down synchronous counter with asynchronous parallel inputs, a counting enable input to enhance synchronous counting capability and an up/down-counting selection input.

Table A1.7 shows the operating table and Table A1.8 the logic table of the counting propagation outputs. Neither table is necessary if using the standard symbols of Figure A1.56, where the function of the input and output pins is as follows:

$\overline{E_A}$ Carry propagation input
$\overline{A}/D$ Up/down counting mode input
T   Input with a double function:
                             - counting: up if G1 = 1 and M2 = 0 and down if G1 = 0 and M2 = 1
                             - propagation enable
$\overline{E_p}$ Control of the parallel information input
Dn(4D)   Parallel data input. The logic value present at this input is introduced in the corresponding flip-flop when C4 = 0

TABLE A1.7

		Inputs			Outputs
$\overline{E_p}$ (C4)	$\overline{A}/D$ (M2)	$\overline{E_A}$ (G1)	T (1,$\overline{2}$-/1,2+) (G3)	(4D)	Q_n
0	X	X	X	0	0
0	X	X	X	1	1
1	0	0	↑	X	Count up
1	1	0	↑	X	Count down
1	X	1	X	X	Counting disable

TABLE A1.8

	Inputs		Contents				Outputs	
$\overline{A}/D$ (M2)	$\overline{E_A}$ (G1)	T (1,2-/1,$\overline{2}$+) (G3)	Q_3	Q_2	Q_1	Q_0	Max/Min	Q_n
1	1	X	1	1	1	1	0	1
0	1	X	1	1	1	1	1	1
0	0	⎓⎍	1	1	1	1	⎍	⎓
0	1	X	0	0	0	0	0	1
1	1	X	0	0	0	0	1	1
1	0	⎓⎍	0	0	0	0	⎍	⎓

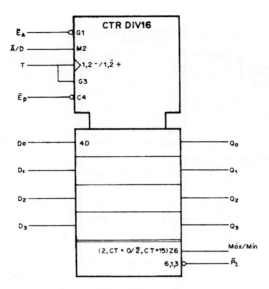

FIGURE A1.56 *Standard symbol of 74191 synchronous counter.*

Max/Min
[2,CT = 0/$\bar{2}$,CT = 15]Z6
 — Output which indicates that the content of the counter has reached the maximum of its counting capacity. This output is set to 1 if M2 = 1 (2) and the content of the counter is 0 or if M2 = 0 ($\bar{2}$) and the content of the counter is 15. This output is connected to the function $\overline{P_I}$

$\overline{P_I}$ (6, 1, 3)
 — Output which stays at 0 if Max/Min = 1, G1 = 1 and G3 = 1

Shift registers

Shift registers are synchronous functional blocks with a multitude of options. Figure A1.57 shows the general symbol of a shift register with a serial input and a serial output. Figure A1.58a shows an example consisting of integrated circuit 7491 which is an 8-bit shift register (SRG8) with:

- a right shift positive-edge-triggered pulse input (C1→);
- a serial information input formed by an AND function of two input pins;
- a serial information output and its inverse.

Figure A1.58a can be simplified to Figure A1.58b. Such simplification is possible because SRG8 indicates that it is an 8-bit shift register.

However, shift registers are used in particular for the serial transmission of data generated in parallel or for executing the inverse operation, i.e. to

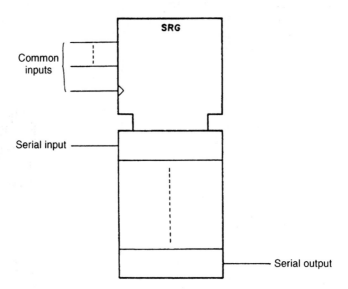

FIGURE A1.57 *General standard symbol of a shift register.*

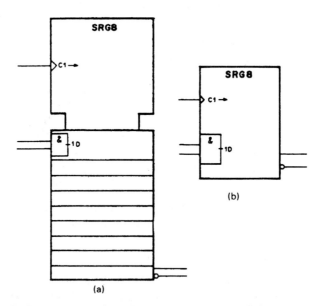

FIGURE A1.58 *Standard symbol of 7491 shift register.*

retrieve in parallel information transmitted in series. Figure A1.59 shows the general standard symbol of a shift register which can perform both functions and, therefore, has serial and parallel inputs, serial and parallel outputs and common control inputs.

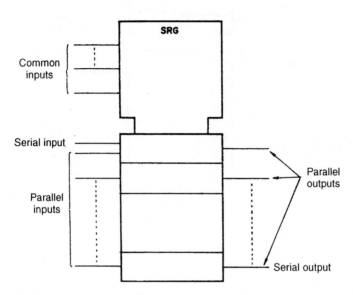

FIGURE A1.59 *Standard symbol of a shift register with serial and parallel input and output.*

A practical example is the integrated circuit 74LS194, which is a 4-bit bidirectional shift register. Figure A1.60 shows the standard symbol with its terminals which have the following functions:

R	Asynchronous reset of the register flip-flops when at 0
S_0, S_1 (M0/3)	Operating mode selection inputs
M_0	Shift disable
M_1	Right shift
M_2	Left shift
M_3	Parallel input
C4/1→/2←	Rising- or positive-edge-triggered pulse input whose action depends on the state of the operating mode selection inputs M0/3:
	• if M0/3 = 0, the applied pulses have no effect
	• if M0/3 = 1, the pulses shift the data to the right (1→)
	• if M0/3 = 2, the pulses shift the data to the left (2←)
	• if M0/3 = 3, the parallel input is enabled
1,4D	Right serial data input. This action is enabled if inputs S_1 and S_0 are at 0 and 1, respectively, equivalent to 1_{10} (M0/3 = 1)

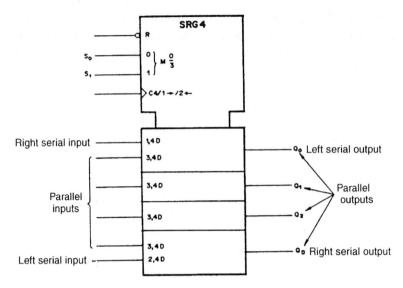

FIGURE A1.60 *Standard symbol of 74LS194 shift register.*

2,4D   Left serial data input. This action is enabled if inputs S_1 and S_0 are at 1 and 0, respectively, equivalent to 2_{10} (M0/3 = 2)

3,4D   Parallel data input. This action is enabled if inputs S_1 and S_0 are both at state 1, equivalent to 3_{10} (M0/3 = 3)

APPENDIX 2

Real logic controller

Figure A2.1 shows a logic controller developed at McGill University. This logic controller is an example of the situation mentioned in chapter 1 in relation to the infinite number of possible ways to implement a logic controller.

Figure A2.2 shows the format of the instructions of the passive PROM memory. The control variable $C/\overline{P}$ of the counter is generated by combining the selected input variable with bits 14 and 15 of the instruction.

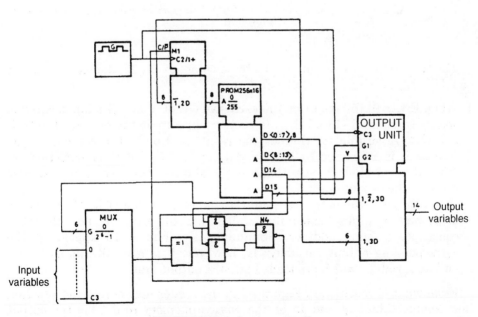

FIGURE A2.1 *Logic controller*.

353

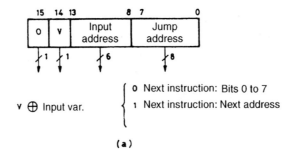

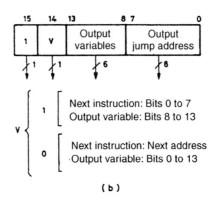

FIGURE A2.2 *Instruction format of the PROM memory of the logic controller of Figure A2.1.*

The two vectors are distinguished by the state of bit 15, denoted as output enable ($EN_\emptyset$), acting in the following way:

1. When $EN_\emptyset = 0$, the result of the exclusive OR of V and the input variable selected by bits 8–13 of the instruction appears at the output of NAND gate N4. According to whether the result is 0 or 1, the content of the counter is incremented by 1 or the data of bits 0–7 of the instruction are entered in parallel, constituting a jump address. The level 0 of $EN_\emptyset$ also disables the output unit.
2. When $EN_\emptyset = 1$, the output of gate N4 equals the value of V. If this value is 1, bits 0–7 are loaded in parallel into the counter, and therefore constitute a jump address. Simultaneously, bits 8–13 appear at the corresponding output variables. If, however, V is a 0, all bits 0–13 are output variables which are loaded into the output unit.

From what is specified in Figure A2.2, the reader may obtain the function that combines bits 14 and 15 of the passive memory to disable the output flip-flops 0–7 and 8–13.

We can conclude that this controller has been designed to minimize the length of the instruction by combining the jump address with a subset of the output variables.

APPENDIX 3

Commercial programmable logic devices

A3.1 Introduction

In chapter 3, logic controllers were implemented using several integrated PLDs. A brief description of each of them is given below. For a more detailed treatment, the reader should refer to the bibliography at the end of this appendix.

A3.2 Programmable logic device PLS100

The PLS100 is a programmable logic array (PLA) of 16 input variables, 48 logic products and 8 output variables. It is implemented in bipolar technology and it is programmed by means of nickel-chromium precision fuses. Figure A3.1 shows the corresponding schema.

The macrocell of each output is very simple because it is formed by:

- A two-input exclusive OR gate (Figure A3.1) with an input that can be set to 0 or 1 by fusing the corresponding fuse. The other input is connected to the output of the corresponding OR gate of the array.
- A tri-state follower gate whose input is connected to the output of the corresponding exclusive OR gate (Figure A3.1). This gate has a control input of the tri-state output; this input is connected to the external pin CE (chip enable). When this input is 1, the output is in the third state. Its purpose is to allow the direct external connection of several arrays to increase its capacity.

Figure A3.2 shows the two chip package outlines, the dual in-line (DIL) and the leadless chip carrier (LCC).

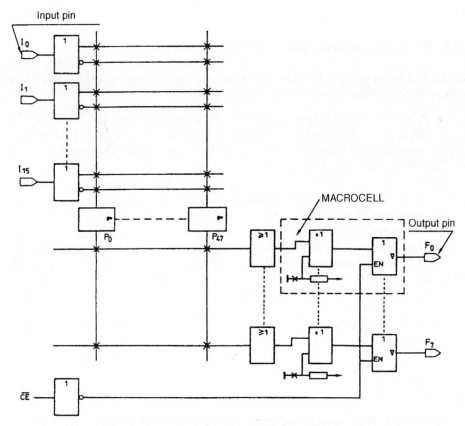

FIGURE A3.1 *Logic circuit of PLS100 PLD.*

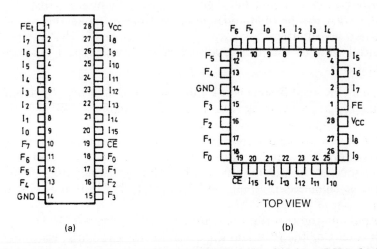

(a) (b)

FIGURE A3.2 *Package outlines of PLS100 PLD.* (a) *Dual in-line (DIL).* (b) *Leadless chip carrier (LCC).*

A3.3 Programmable logic device PLS155

The PLS 155 is a bipolar PLD programmed by means of nickel-cadmium fuses with the following elements (Figure A3.3):

- A programmable logic array with 32 AND gates and 21 OR gates. The AND gate array has 4 specific input variables I_0-I_3 and 13 feedback variables, 12 of which come from the output macrocells and 1 that is directly fed back internally.
- Eight type 1 macrocells whose inputs are connected to the outputs of an identical number of OR gates.
- Two type 2 macrocells with a register formed by two synchronous flip-flops.
- One array of 11 AND gates. It is called a control array because its outputs are connected to the control variables of the operating mode of the macrocells. This array has the same input and feedback variables as the AND gate array which forms part of the programmable logic array (PLA).
- One independent control array of two AND gates with an input variable ($\overline{OE}$) which controls the tri-state outputs of type 2 macrocells.

The type 1 macrocell is shown in Figure A3.4 and is almost identical to the one in PLD PLS100 (Figure A3.1) except that:

- the tri-state control variable EN is connected to the output of an AND gate of the control array; and
- the output of the macrocell is connected to both AND gate arrays through inverter and follower amplifiers. Combining this characteristic with the previous one, it is possible to make the external pin either an input or an output, at the designer's discretion. In the initial state of the device, all external pins of type 1 macrocells are inputs.

The type 2 macrocell is shown in Figure A3.5 and is formed by a register of two rising-edge-triggered J–K flip-flops with common set (S) and reset (R) inputs connected to a pair of OR gates.

Inputs J and K are connected to the input and output, respectively, of a pair of inverters N1 and N2 with a tri-state output controlled by an EN variable connected to the output of the AND control gate P_c through a programmable connection. When the EN input is 1, flip-flop J–K becomes a D flip-flop.

The outputs of the two flip-flops are fed back to the two AND arrays (the one which is part of the PLA and the control array) and are also connected to a pair of follower gates N3 and N4 with tri-state outputs connected to external pins (Figure A3.5). The tri-state enable input (EN) of these gates is connected

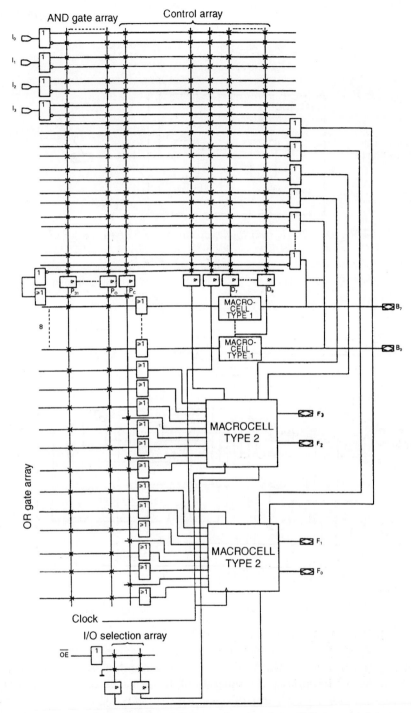

FIGURE A3.3 *Logic circuit of PLS155 PLD*.

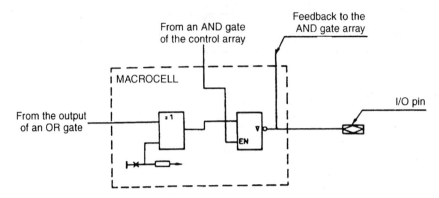

FIGURE A3.4 *Logic circuit of type 1 macrocell of PLS155 PLD.*

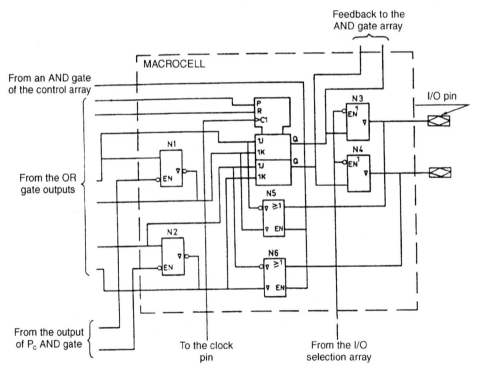

FIGURE A3.5 *Logic circuit of type 2 macrocell of PLS155 PLD.*

to an $\overline{OE}$ pin through a two-AND gate programmable array, denoted I/O selection array, which allows the outputs of those gates to be set to a static zero or to be made a function of $\overline{OE}$. In this way, the above-mentioned external pins can become inputs or outputs. To become inputs they are connected to a pair of follower and inverter tri-state gates N5 and N6 whose outputs are connected to inputs J and K of the flip-flops (Figure A3.5). In this way we have synchronized inputs through the two J–K flip-flops.

Figure A3.6 shows the two package outlines available for PLS155: the dual in-line (DIL) and the plastic leadless chip carrier (LCC).

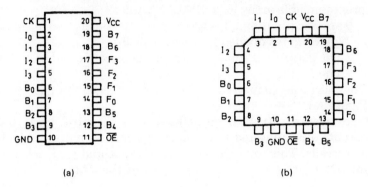

FIGURE A3.6 *Package outlines of PLS155 PLD.* (a) *Dual in-line (DIL).* (b) *Leadless chip carrier (LCC).*

A3.4 Programmable logic device PLS157

This PLD is similar to the PLS155 except that:

1. It has six type 1 macrocells instead of eight.
2. It has one type 2 macrocell (Figure A3.5) and another one with four flip-flops instead of two.

Figure A3.7 shows the two package outlines available for the PLS157: the dual in-line (DIL) and the leadless chip carrier (LCC).

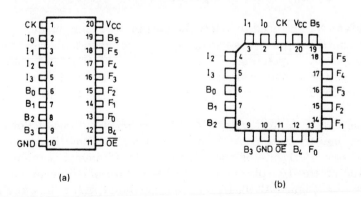

FIGURE A3.7 *Package outlines of PLS157 PLD.* (a) *Dual in-line (DIL).* (b) *Leadless chip carrier (LCC).*

A3.5 Programmable logic device 5C031

The 5C031 PLD has the following elements (Figure A3.8):

- An AND programmable logic array (the manufacturers usually call it 'programmable array logic (PAL)') with:
  9 specific input pins I_0-I_8
  72 AND gates divided in 8 groups of 9
  8 OR gates of 9 inputs, each of them connected to the output of one AND gate.
- One macrocell connected to the output of each OR gate, as shown in Figure A3.9.
- One array of 10 AND gates known as a control array because its outputs are connected to the operating mode control variables of the macrocells. This array has the same input and feedback variables as the AND gate array which forms part of the PAL.

Each macrocell (Figure A3.9) consists of one edge-sensitive D flip-flop with one synchronous set (1S), one asynchronous reset (R) and one set of tri-state follower gates N1–N8.

All the synchronous set inputs (1S) are connected to the output of one AND gate of the control array. The same happens with the reset (R) inputs of all the flip-flops.

Gates N1–N3 determine how feedback is carried out. Each has an enable variable (EN) to control the tri-state output and the value of these EN variables is selected by programmable MOS transistors (represented by X in Figure A3.9). The outputs of N1, N2 and N3 are connected and fed back to the AND gate array. If input EN of N1 is set to level 1 (activating the corresponding MOS transistor), the output of the OR gate of the programmable array is fed back. If input EN of N2 is set to 1, the output of the D flip-flop is fed back. Finally, if input EN of N3 is set to 1, the I/O pin state is fed back.

Gates N4–N7 are used to select the output operating mode. Each has one tri-state control input whose logic level is programmable by means of a MOS transistor. Depending on the logic value of the EN inputs of N4 to N7, the common output connection equals the output of the corresponding OR gate of the PAL, its inverse, the output Q or the output $\bar{Q}$ of the D flip-flop.

The common output of gates N4–N7 is connected to the follower gate N8 which has one tri-state control enable input EN which is connected to the output of one AND gate of the control array. If such EN input is a 1, the logic value of the I/O pin equals the logic state of the common output of gates N4–N7 (Figure A3.9). If, on the other hand, EN is a 0, the I/O pin is converted into an input and its data can be transmitted to the AND gate array, through N3.

APPENDIX 3

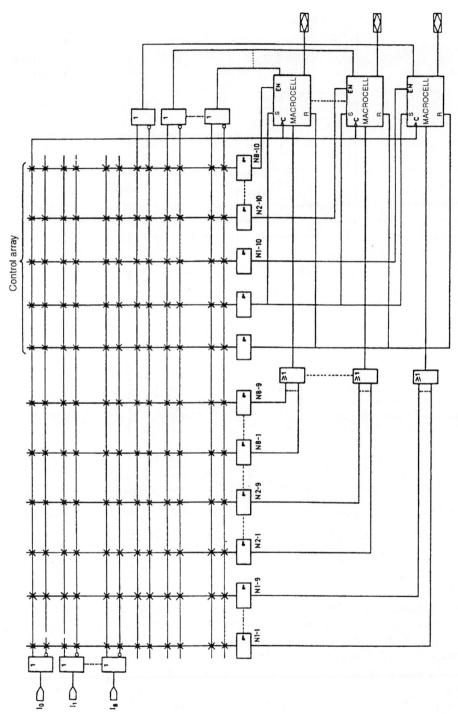

FIGURE A3.8 *Logic circuit of 5C031 PLD.*

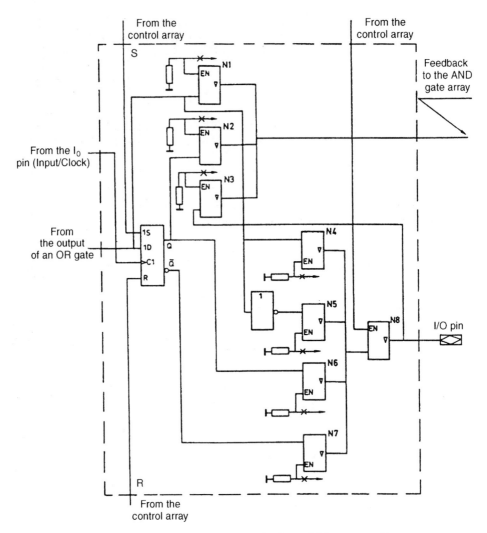

FIGURE A3.9 *Logic circuit of 5C031 PLD macrocell.*

Because the EN input of N8 is connected to the output of the AND gate array, the output of N8 can constitute a bus where, at times, the data from the PLD itself or from an external source can appear.

Table A3.1 shows the different combinations of the states of the EN inputs of gates N1–N8. Every combination is a different way of implementing the input, output and feedback, and constitutes one configuration of the PLD.

Figure A3.10 shows the package outline of PLD 5C031.

APPENDIX 3

TABLE A3.1

Logic value of the EN input of gates N1 to N8	Feedback type	Output type	PLD Configuration
$EN_{N1}=EN_{N4}=EN_{N8}=1$ The rest$=0$	Combinational	Combinational	Combinational output and Combinational feedback (COCF)
$EN_{N1}=EN_{N5}=EN_{N8}=1$ The rest$=0$	Combinational	Combinational and inverted	Combinational output and Combinational feedback (COCF)
$EN_{N1}=EN_{N6}=EN_{N8}=1$ The rest$=0$	Combinational	Registered	Registered output and Combinational feedback (ROCF)
$EN_{N1}=EN_{N7}=EN_{N8}=1$ The rest$=0$	Registered	Registered and inverted	Registered output and Combinational feedback (ROCF)
$EN_{N2}=EN_{N4}=EN_{N8}=1$ The rest$=0$	Registered	Combinational	Combinational output and Registered feedback (CORF)
$EN_{N2}=EN_{N5}=EN_{N8}=1$ The rest$=0$	Registered	Combinational and inverted	Combinational output and Registered feedback (CORF)
$EN_{N2}=EN_{N6}=EN_{N8}=1$ The rest$=0$	Registered	Registered	Registered output and Registered feedback (RORF)
$EN_{N2}=EN_{N7}=EN_{N8}=1$ The rest$=0$	Registered	Registered and inverted	Registered output and Registered feedback (RORF)
$EN_{N3}=1$ The rest$=0$	External input	Do not exist	No output (NOIF) and external input

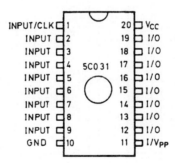

FIGURE A3.10 *Package outline of 5C031 PLD.*

A3.6 Programmable logic device 5C060

The 5C060 is shown in Figure A3.11 and consists of:

- One AND-programmable logic array (PAL) with:
  4 specific input pins I_1–I_4
  128 AND gates divided into 16 groups of 8

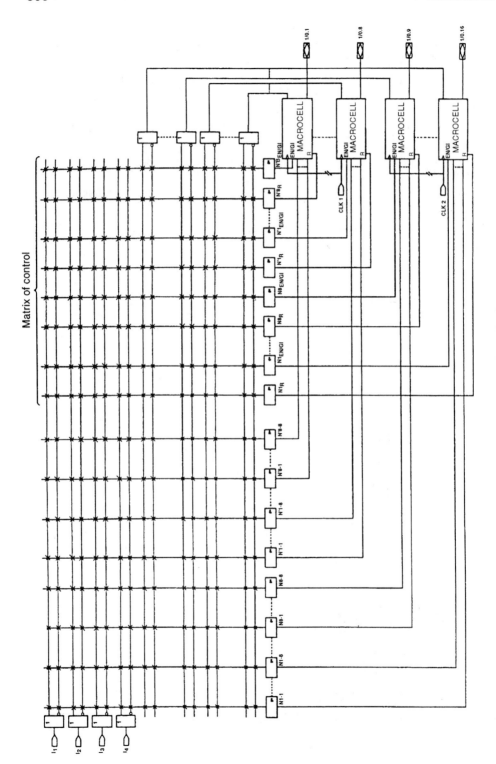

FIGURE A3.11 *Logic circuit of 5C060 PLD.*

APPENDIX 3

1 set of OR gates contained inside the output macrocells. Each OR gate is part of the programmable combinational circuit of the macrocell
- One set of 16 macrocells whose diagram is shown in Figure A3.12.
- One array of 16 AND gates designated as a control array because its outputs are connected to the operating mode control variables of the macrocells. This array has the same input and feedback variables as the PAL.

The sixteen macrocells are identical and are divided into two groups of eight, each with its own clock input.

Figure A3.12 shows the functional diagram of each macrocell. The basic element of each is the universal flip-flop which has one rising-edge-triggered cell and one programmable combinational circuit with MOS floating gate transistors. Depending on how it is programmed, the cell becomes a D, T, J–K or S–R flip-flop.

When the cell becomes a D or T flip-flop the outputs of the 8 AND gates are combined through an OR gate whose output is connected to the (D or T) flip-flop input. When the cell becomes a J–K or S–R flip-flop, it is possible to program N AND gates connected to an OR gate which is in turn connected to

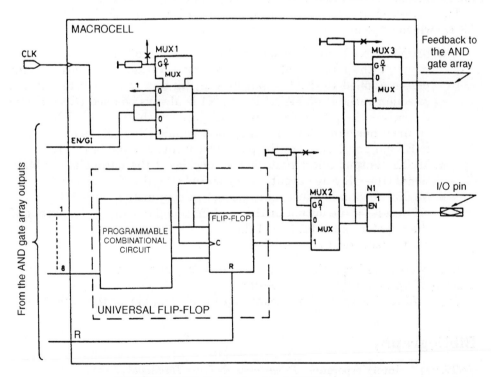

FIGURE A3.12 *Logic circuit of 5C060 macrocell.*

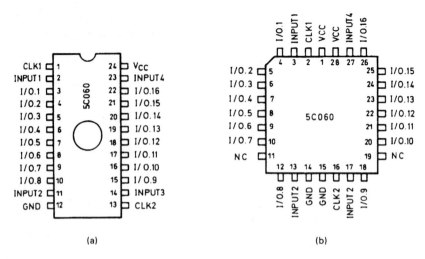

FIGURE A3.13 *Package outlines of 5C060 PLD.* (a) *Dual in-line (DIL).* (b) *Leadless chip carrier (LCC).*

the input J or R, and $8N$ AND gates connected to another OR gate which is in turn connected to the input K or S, where N can vary from 1 to 7.

The macrocell also has three multiplexers MUX1, MUX2, MUX3 (Figure A3.12) which implement the following functions:

- MUX1 (EN/GI multiplexer) allows the state of the enable control input of the tri-state of the output gate N1 and the source of the pulses applied to the C input of the flip-flop to be selected. If the selection variable of MUX1 is a 0, N1 is always enabled (EN = 1) and C receives the pulses from the AND gate which generates EN/GI. If, on the other hand, MUX1 is a 1, the variable EN of N1 is made equal to EN/GI and C receives the pulses from the external pin CLK.
- MUX2 (output select multiplexer) decides if the output is combinational (the OR gate output) or registered (the flip-flop output).
- MUX3 (feedback select multiplexer) in conjunction with MUX2 and N1 decides if the output of the combinational circuit, or the output of the register, is fed back or if the external data present at the I/O pin are transmitted to the AND gate array.

It is worth noticing the great flexibility of this circuit and the large number of options offered to the designer.

Figure A3.13 shows the two package outlines available for the 5C060.

Bibliography

[INTE 94]   Intel Corporation, *Programmable Logic Handbook*, 1994.
[PHIL 94]   Philips, *Programmable Logic Devices*, IC13 data handbook, 1994.

INDEX

Active memory, 48, 279
Address, 47
   dependency relationship, 327
Analog variable, 285, 288, 297
AND
   dependency relationship, 323
   gate, 53, 59
   gate array, 47, 51, 55, 64
   instruction, 160, 170, 173, 207, 248
   programmable gate array, 55
AND-OR programmable logic array, 89
Arithmetic instruction, 291, 300
Assembler program, 104
Asynchronous
   counter, 344
   logic controller, 8
   reset, 88

BCD natural, 28
Binary natural, 28
Block transfer instruction, 293
Buried state register, 70, 120

Bus, 30, 38
   global, 83
   local, 83

CAD, 91, 94
   tools, 94
Canonical product, 51, 53
Cell, 47
Central Processing Unit, 279
Change instruction, 140
Circuit
   capture, 92
   simulation, 95

Clock, 9, 13, 331
Closed-loop control system, 3
CMOS, 48, 89
Code conversion instruction, 302
Combinational
   PLD, 47
   programmable logic device, 47
   system, 7, 47
   logic controller, 4, 92, 94
Combination of dependency
   relationships, 5, 329, 335
Communications
   interfaces, 285
   processor, 312
   unit, 312
Comparator, 31
Complete
   device, 47
   universal combinational PLD, 47
Computer, 276, 278
   aided design, 91, 94
Configurable integrated circuit, 45
Control
   dependency relationship, 332
   instruction, 165, 223, 253
Controller
   closed-loop, 3
   logic, 3
   open-loop, 3
Copyright, 45
Cost reduction, 91
Counter, 12, 344
   asynchronous, 344
   synchronous, 346
   up counter, 164, 346
   up-down counter, 164, 348

Counting instruction, 162, 172, 174, 217, 251
Coupling module, 91
CPU, 279
Custom integrated circuit, 27

D type flip-flop, 31, 60, 63, 85, 120
Data
  memory, 284
  module, 296
Decoder, 30, 52
Delay, 65
Demultiplexer, 78
Dependency relationship
  address, 327
  AND, 323
  combination of, 329, 335
  control, 332
  enable/disable, 326
  interconnection, 327
  negation, 325
  operating mode, 326, 334
  OR, 324
  reset and set, 332
Design workstation, 91
Digital electronics, 4, 53
Disable dependency relationship, 326

Edge-triggered, 15, 16
  flip-flop, 60, 85, 120
  register, 10, 120
EEPROM memory, 50, 125
EPROM memory, 11, 50, 125
Enable
  dependency relationship, 326
  input (EN), 31, 326
  input (G1), 31, 326
Equality instruction, 139
Exclusive OR, 60
Expander NAND gate, 79
External output variable, 158

Fan-out, 81
Fault, 95
Feedback path, 60, 68, 69, 71
Fixed
  allocation PLD, 69
  connection, 51, 74
  read only memory, 48
FLASH memory, 50, 125
Flip-flop, 9, 31, 60, 63, 85
Floating gate MOS transistor, 89

Flow diagram, 94
Flowchart, 18, 24
Foldback
  NAND gate array, 84, 89
  connection, 89
Follower gate, 70
Functional
  block, 313
  modules, 296
Fuse, 88, 89
  map file, 104

Gate
  AND, 53, 59
  OR, 60
  tri-state, 60
Global array, 80
GRAFCET, 176
Graphic design software, 92

Hazard, 8, 9
High-level language, 94, 104
High speed timer, 251

Incomplete universal combinational PLD, 52
Information display capability, 307
Input
  modularity, 30
  symbol, 316
  unit, 31, 125
  variable, 158, 206, 245
Instruction
  arithmetic, 291, 300
  block transfer, 293
  change, 140
  conditional disabling, 134
  conditional jump, 223
  conditional operating, 134
  control, 165, 223, 253
  counting, 162, 172, 174, 217, 251
  equality, 139
  input, 158
  jump, 131, 223, 294
  list language, 158
  load, 128, 247
  logic, 137
  return, 294
  shift register, 293, 303
  store, 128
  timing, 162, 170, 174, 211, 249
  transfer, 290

INDEX

Integrated circuit, 41, 45
  semi-custom, 8, 27
Integration density, 89
Interconnection dependency relationship, 327
Interface, 283
  analog to digital, 285
  communication, 285
  digital to analog, 285
Internal
  output variables, 158, 206, 245
  state, 15, 17
  state register, 12
Inverted
  logic AND instruction, 137
  logic OR instruction, 138
Inverter, 62
  gate, 70

Jump
  address, 37
  instruction, 131, 223, 294

Karnaugh table, 6, 53
Keep instruction, 253

Ladder diagram, 167
Large scale integration, 45, 331
Latch register, 11, 13
Leadless chip carrier, 356, 361, 368
Level change, 16
Local array, 80
Logic
  diagram, 92
  equation, 94
  equation minimization, 94
  function, 36, 320
  gate symbol, 320
  unit, 125
Logic controller, 3
  asynchronous, 8
  combinational, 4, 92
  modular, 15, 30
  non-modular, 14, 25
  programmable, 26
  semi-modular, 15, 39
  sequential, 7, 92, 94
  synchronous, 7, 9
  wired, 27, 38
LSI, 45, 331

Macrocell, 62
  operation mode, 63

Mealy sequential system, 61
Medium scale integration, 45, 331
Memory
  EEPROM, 50, 125
  EPROM, 11, 50, 125
  FLASH, 50, 125
  position, 48, 50
  PROM, 11, 26, 50, 125
  Random access, 47
  ROM, 11, 25, 50, 125
Microcomputer, 281
Microelectronics, 30
Microinstruction, 51
Microprocessor, 281
Minicomputer, 91
Minimal non-canonical expression, 53
Minimization program, 94
Modular logic controller, 15, 30
Modularity, 8, 9, 30
  Input, 30
  Output, 30
MOS floating gate transistor, 88, 89
MSI, 45, 331
Multiple feedback, 68
Multiplexer, 61, 63, 65

NAND gate, 7
Negation dependency relationship, 325
Noise immunity, 45
Non-modular logic controller, 25
Non-programmable array, 51
NOR gate, 7

One hot encoding, 146, 150
One-time programming, 89
Open-loop control system, 3
Operating mode relationship, 326, 334
Operations with words, 292
Operation
  code, 39, 127
  mode variable, 38
OR
  dependency relationship, 324
  gate, 60
  gate array, 51, 52, 56
  instruction, 159, 169, 173, 207, 248
Organization module, 296
Output
  instructions, 158, 207, 247
  modularity, 30
  symbol, 316
  unit, 31, 125

PAL, 11, 26, 57
PAL-based
    PLD, 62
    PLS, 59
Parallel register, 339
Passive memory, 48, 279
Peripheral, 281
    unit, 310
PIA, 59, 69, 79
PLA, 11, 26, 55
PLC, 39, 125
    instruction, 126, 158
    programming languages, 157
PLD, 41, 45
    advanced PAL-based, 68
    expander product term array, 79
    fixed allocation, 69
    logic product steering, 73
    logic sum-of-product steering, 79
    multiple array, 80
    multiple product term allocation, 78
    non-segmented, 69
    programmer, 91
    segmented, 69, 80
    variable allocation, 69
PLS, 59
Power dissipation, 45, 89
Printed circuit, 45, 91
Process controller, 3
Program
    memory, 284
    module, 296
Programmable
    array logic (PAL), 11, 26, 57
    combinational circuit, 38, 47
    comparator, 47
    connection, 52
    element, 63, 89, 95
    interconnect array, 59
    interconnect matrix, 59
    logic array (PLA), 11, 26, 55
    logic controller (PLC), 26, 39, 125
    logic device, 30, 41, 45
    logic sequencer, 59
    read only memory, 50
    universal combinational system (PUCS), 25
Programming file, 95
Programming languages
    function diagram, 172
    GRAFCET, 176
    instruction list, 158
    ladder diagram, 167
    relay diagram, 167
Programming
    module, 95
    unit, 306
PROM, 11, 26, 50, 125
Propagation time, 45
Pulse generator, 9, 86, 331

RAM, 47
Random Access Memory, 47
Read only memory, 48, 50
Read/write memory, 48
Reading operation, 48
Register, 9
    parallel, 339
    shift, 349
Relay diagram, 167
Reliability, 45, 91
Reprogrammable read only memory, 50, 89
Reset
    dependency relationship, 332
    input, 60
    instruction, 140, 220, 253
ROM, 11, 26, 50, 125
RPROM, 11, 26, 50, 89, 125
RS 232-C, 91, 310
R-S flip-flop, 153, 183, 220, 253

Selection input, 31, 63
Self-retention function, 220
Semi-custom integrated circuit, 8, 27
Semi-modular logic controller, 39
Semi-modularity, 39
Sensor, 6
Sequence of decisions, 37
Sequential
    logic controller, 7, 92, 94
    PLD, 58
Set
    dependency relationship, 332
    input, 60
    instruction, 140, 220, 253
Shift
    operations, 293, 303
    register, 349
    register instruction, 293, 303
Simulation program, 95
Small scale integration, 45
Special instructions, 293

S-R flip-flop, 153, 183, 220, 253
  instruction, 253
SSI, 45
Standard
  logic symbol, 9, 126
  off-the-shelf integrated circuit, 41, 45
  serial connection, 91, 310
State
  code assignment, 105
  diagram, 15, 94
  transitions, 94
  vector, 111
Steering circuit, 74
Structured programming, 295
Sum-of-products, 55, 127, 128
  steering, 74
Synchronization
  flip-flop, 71
  register, 10, 11, 13, 119
Synchronous
  register, 9
  sequential system, 7, 9, 12
Synchronous flip-flop, 9, 59
Synchronous logic controller, 7, 9
  modular, 15, 30
  non-modular, 14, 25
  semi-modular, 15, 39
Syntactic analysis, 104

Temporary memory, 128
Third state, 40
Timing instruction, 162, 170, 174, 211, 249

non stored pulse triggering (SP), 211
stored pulse triggering (SE), 212
stored switch-on delay (SS), 216
switch-off delay (SF), 216
switch-on delay (SD), 213
Timing variable, 162, 211
Transfer instruction, 290, 299
Transformation operations, 292
Transistor-transistorlow power Schottky
  technology, (LSTTL), 89
Transition
  capacity, 17, 145
  graph, 149
Tri-state, 50, 326
  gate, 60
Truth table, 4, 94, 96, 99
Two-dimensional array, 51

Universal
  combinational PLDs, 47
  gate PLD, 84

Variable
  allocation PLD, 69, 72
  identification, 158, 167, 172, 206, 245
Vector, 37, 39, 126
Very large scale integration, 45, 331
VLSI, 45, 331

Wired logic controller, 27, 38
Workstation, 91
Writing operation, 48